Uxpanapa:
Agricultural Development in the Mexican Tropics

Pergamon Titles of Related Interest

Chou/Harmon Critical Food Issues of the Eighties
Goodman/Love Food Projects
Miljan Food and Agriculture in Global Perspective

Related Journals*

Habitat International
Socio-Economic Planning Sciences
World Development

*Free specimen copies available upon request.

PERGAMON POLICY STUDIES ON INTERNATIONAL DEVELOPMENT

Uxpanapa: Agricultural Development in the Mexican Tropics

Peter T. Ewell
Thomas T. Poleman

Pergamon Press
NEW YORK • OXFORD • TORONTO • SYDNEY • FRANKFURT • PARIS

Pergamon Press Offices:

U.S.A.	Pergamon Press Inc., Maxwell House, Fairview Park, Elmsford, New York 10523, U.S.A.
U.K.	Pergamon Press Ltd., Headington Hill Hall, Oxford OX3 OBW, England
CANADA	Pergamon of Canada, Ltd., Suite 104, 150 Consumers Road, Willowdale, Ontario M2J 1P9, Canada
AUSTRALIA	Pergamon Press (Aust.) Pty. Ltd., P.O. Box 544, Potts Point, NSW 2011, Australia
FRANCE	Pergamon Press SARL, 24 rue des Ecoles, 75240 Paris, Cedex 05, France
FEDERAL REPUBLIC OF GERMANY	Pergamon Press GmbH, Hammerweg 6, Postfach 1305, 6242 Kronberg/Taunus, Federal Republic of Germany

Library of Congress Cataloging in Publication Data

Ewell, Peter T 1947-
Uxpanapa.

(Pergamon policy studies)
Bibliography: p.
Includes index.
1. Agriculture and state—Mexico—Papaloapan Valley.
2. Agriculture—Mexico—Papaloapan Valley. I. Poleman, Thomas T., joint author. II. Title.
HD17895.P3E93 1980 338.1'0972'6 80-12208
ISBN 0-08-025967-7

Printed in the United States of America

Este estudio esta dedicado

a la memoria del

Ing. Jorge L. Tamayo

y del

Ing. Guillermo Hernandez Castro,

Quienes ofrendaron sus vidas

al sueno de Uxpanapa.

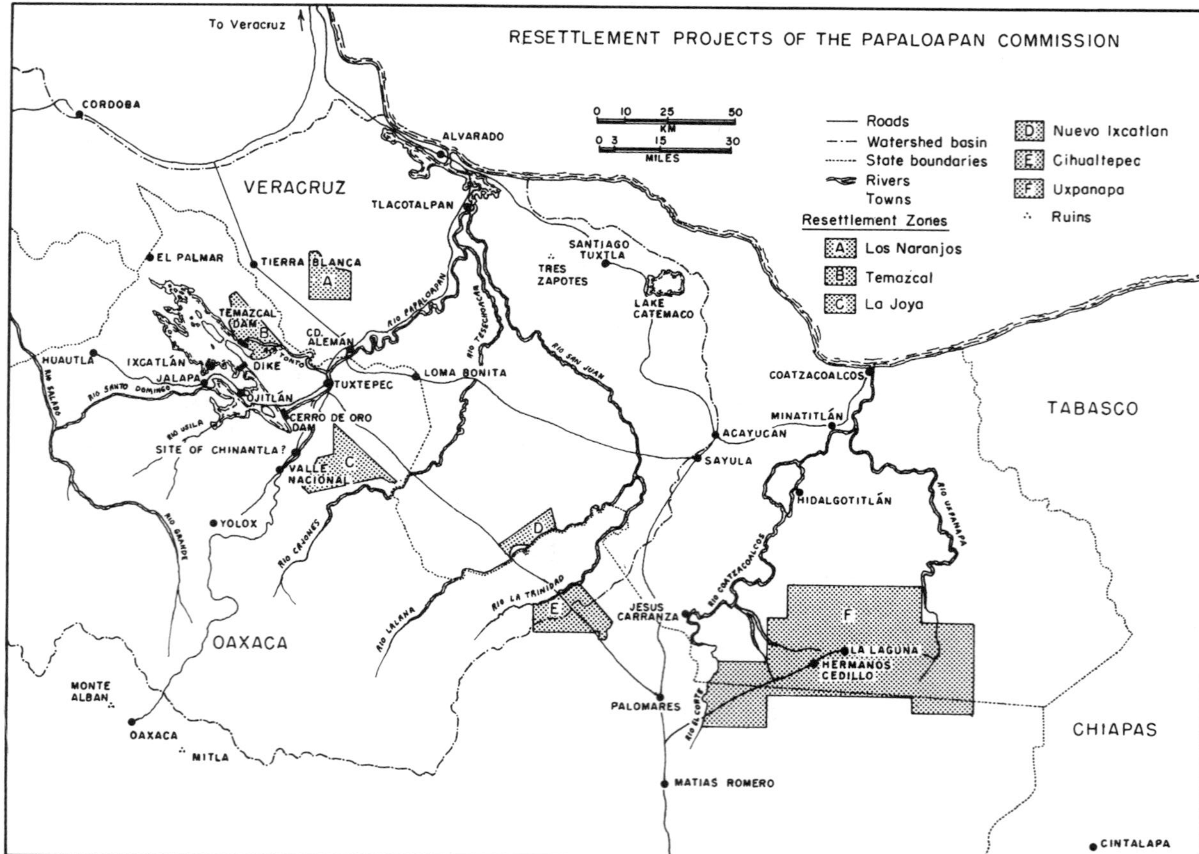

RESETTLEMENT PROJECTS OF THE PAPALOAPAN COMMISSION

Contents

Preface

For me the coastal lowlands of Southeastern Mexico will always have a mystical appeal. Whether approached by road winding down from the altiplano or from the sea at Veracruz, the endless green plain affords the same vista of boundless beauty, opportunity, and challenge. Surely it is the promised land, and surely the gods watching over it dwell atop that snow-capped peak - Orizaba - visible every now and then above the clouds. No fools they: it's cool up there.

It's usually hot down below and was especially so 25 years ago when a group of North American students was ushered into the office of the Vocal Ejecutivo of the Comision del Papaloapan. Power failures were frequent in those days, but the absence of air-conditioning did nothing to dampen the enthusiasm of our host. No one who met Ing. Raul Sandoval Landazuri could not but feel they were in the presence of greatness, and through his words we were made to see the vision of Mexico's tropics at long last developed. And what a vision it was: of disease eliminated, great rivers controlled, highways cutting through the jungle, and the jungle giving way to a modern agriculture capable of feeding the country's rapidly growing population. If it was also somewhat naive, it was a naivete worthy of the aspirations of a great nation.

Thanks to Ing. Sandoval's encouragement and tolerance, I was able to write my doctoral dissertation and publish a small book on the formative years of the Papaloapan Project (Poleman, 1976). Thereafter, my work took me to Africa and Asia, and if I thought at all about Mexico I suppose it was in the context of most governmentally-sponsored tropical development schemes failing to attain their objectives.

Thus my pleasure upon returning in early 1976 was all the greater for the Lower Papaloapan Basin had been utterly transformed. No longer isolated, it had become the home of a booming agriculture. Where once stood mile after mile of untouched bush were now tidy orchards and planted pasture. Sugar - and new sugar mills - were everywhere. Neatly painted houses made of cinder block, the blue glow of a television emitting from each, replaced thatched huts. On the Tinajas-Tuxtepec highway, one saw more traffic in 15 minutes than used to pass in 24 hours. And Tuxtepec, once best described as a "sea of mud," had become a modern metropolis, obviously thriving.

The changes extended to the Papaloapan Commission. When created in 1947, it was the first regional development agency in tropical Mexico. A decade later, its budget had been slashed and bureaucratic sclerosis was all too evident. But by 1976, dynamism again prevailed.

In the distinguished geographer, historian, and administrator, Ing. Jorge L. Tamayo, the Commission had a Vocal Ejecutivo worthy of Ing. Sandoval's dreams; and as its Vocal Secretario, it was fortunate to have a man who had devoted his professional life to the Papaloapan Basin and its development. I first met Ing. Guillermo Hernandez Castro when, as resident at the Los Naranjos zone, he was simultaneously grappling with the problems of Mazatec resettlement, an irrigation trial, and a scheme for agricultural intensification. All three projects ultimately failed, but he was not unduly concerned: developing a new area is very much a matter of trial and error.

"You must see Uxpanapa," Ing. Hernandez said. "There, in one of the most isolated corners of the country, we are applying the lessons of Los Naranjos. Soon 15 thousand Chinantecs will be resettled and will immediately begin farming on a commercial scale. Send one of your students to record what happens."

This study is our response to that invitation.

The project of Uxpanapa has two principal goals. The first is to help meet pressing national needs for food by introducing intensive cropping into a large area of the rainy tropics. The second is to compensate a group of Indians who are being displaced by a flood-control reservoir and to facilitate their transformation into modern farmers. A variety of pressures have forced the Papaloapan Commission to move very rapidly, and pursuit of these goals has not been well coordinated. There have been errors and the project has generated controversy. For clearing so large an area of virgin rain forest, the Commission has been accused of "ecocide"; and of "ethnocide" for uprooting over 40 percent of the Chinantec people.

Uxpanapa is still in its initial phases and it will be many years before these and other charges can be fully evaluated. Nevertheless, a detailed account of the project to date is demanded by the recent discovery of vast deposits of petroleum in the Southeast. The slow and careful development of the remaining land resources of its tropics is a luxury not to be allowed Mexico. The exploitation of the oil will require enormous investment and involve the government in new and little understood areas. Centralized planning and cooperative organization are frequently suggested as vehicles for minimizing the waste and injustice that attend rapid change. If we have succeeded in suggesting that tidy answers are few, the pitfalls many, and that creativity and wisdom of the highest order are called for, we will rest content.

T.T.P.

Acknowledgments

This study and the photographs accompanying it are very much Peter Ewell's; my name as junior author appears as a matter of courtesy. It reflects almost two-and-one-half year's work on his part, ten months of which were spent in the field. His visits to Uxpanapa - and mine - were made possible by a generous grant to Cornell University from the Tinker Foundation and I record my appreciation to Miss Martha Muse and Dr. Erika Bruce, President and Program Director, respectively, of the Foundation, for the sympathetic way in which they responded to what must have seemed a series of odd requests. While at Cornell, Mr. Ewell was supported by an assistantship and back-up services provided by the Department of Agricultural Economics. The maps and charts reflect the talent of Joseph Baldwin. Lillian Thomas, my secretary, typed, kept track of the money, and remembered what we were doing. We are indebted to all.

But our greatest debt is to the many people involved at Uxpanapa - both employees of the Commission and private citizens - who welcomed Mr. Ewell as their parents did me: with tolerance, friendship, and a very real desire to be helpful. We would particularly like to mention: (1) the Comision del Papaloapan - Sta. Sara Flores, Ing. Jose Rodriguez Vallejo, Ing. Rogelio Banderas Valencia, Sr. Mario Burguete S., Ing. Humberto Merchand Rojas, Ing. Jose Luis Casar Rodriguez, Ing. Rurico J. Millan Cepeda, Ing. Miguel A. Lope Avila, Ing. Alberto Guido Astorga, Ing. Jesus Gamez Escarcega, Ing. Jesus Sampers Toledo, Ing. Rafael Rosales Marrufo, Lic. Javier Villalobos, Sr. Oscar del Angel, Ing. Roberto Beltran, and the entire staff of the agricultural office in Uxpanapa; (2) the Banco de Credito Rural del Golfo - Lic. Victor Moye Arrivillega and the entire staff in Uxpanapa, also Ing. Santiago Lecou; and (3) the officials and members of the ejidos, in Ojitlan as well as Uxpanapa.

I close on a note of sadness. Just as Ing. Sandoval was killed in an airplane accident at the time I was finishing my thesis in 1956, so a helicopter crash at Uxpanapa in the closing days of 1978 took the lives of Ings. Tamayo and Hernandez. For all they did to help the people of the Southeast - and Mr. Ewell and my other students - I can only say thanks. They will be sorely missed. Their vision will live on.

T.T.P.

List of Photographs

(following page 91)

Glossary of Acronyms

ANAGSA	Aseguradora Nacional Agricola y Ganadera - The National Crop and Cattle Insurance Company.
CCI	Central de Campesinos Independientes - Union of Independent Campesinos, a small peasant union founded in the 1960s.
CNC	Confederacion Nacional Campesina - National Confederation of Campesinos, the official peasant union and constituent element of the PRI party.
CONACYT	Consejo Nacional de Ciencia y Tecnologia - The National Science and Technology Council, the Mexican equivalent of the National Academy of Sciences.
CONASUPO	Compania Nacional de Subsistencias Populares -The National Basic Foods Company. This organization both buys commodities from farmers at official prices and operates a chain of outlets that sell basic foods at subsidized prices.
DAAC	El Departamento de Asuntos Agrarios y Colonizacion - The Department of Agrarian Affairs and Colonization. This organization administered the agrarian reform process and created ejidal colonies in frontier areas. Under the Echeverria administration, it was given ministerial status as the Secretaria de la Reform Agraria.
FIRA	Fideicomiso Instituido en Relacion con la Agricultura - A specialized agency of the central Bank of Mexico, which finances long-term credit in the agricultural sector.

FONAFE	Fondo Nacional de Fomento Ejidal - The National Fund for Agricultural Development, which administered the common funds of ejidos and financed the establishment of industries in the reformed sector. Under the Lopez Portillo administration, the agency was reorganized as the Fideicomiso de Fondos Comunes.
INI	Instituto Nacional Indigenista - The National Indian Affairs Institute.
INIA	Instituto Nacional de Investigaciones Agricolas - The National Agricultural Research Institute.
PARM	Partido Autentico de la Revolucion Mexicana - The Authentic Party of the Mexican Revolution, a minor political party.
PRI	Partido Revolucionario Institucional - The Institutional Revolutionary Party, the single official party of the Mexican government.
PEMEX	Petroleos Mexicanos - The Mexican National Oil Company.
PRONASE	Productora Nacional de Semillas - The National Seed Production Company.
PROQUIVEMEX	Productos Quimicos Vegetales Mexicanos - The Mexican Vegetable Chemical Products Company, an agency set up to buy, process, and market barbasco - a tropical root crop from which steroids used in the production of birth control pills are derived.
SESA	Servicios Ejidales, S.A. - The Ejidal Services Company, a subsidiary of the official Rural Bank that provides custom-machinery service to the ejidal sector.
SICA	Sociedad Regional de Interes Colectivo Agricola - Regional Collective Agriculture Society, an institution created in the 1930s to coordinate mutual investments of collective ejidos in the irrigation districts in the North.
TABAMEX	Tabacos Mexicanos - The Mexican Tobacco Company, which provides credit to growers and markets the product.

1 Modern Agriculture in the Rain Forest

Uxpanapa, Veracruz, is a 1,000-square-mile area of tropical rain forest. The Mexican government is opening it up to resettle over 15,000 Chinantec Indians who will be flooded out by the construction of the Cerro de Oro dam in northern Oaxaca, 150 miles away to the northwest. Four of us were riding in the cab of a pick-up; the chief agronomist for the district, the manager of the local branch of the National Rural Credit Bank, an inspector from the National Crop Insurance Company, and myself. The road was rutted from four months of almost continuous downpour. The rain had let up earlier in the morning, and the sun was pulling the moisture up from the ground so strongly that we were all soaking wet.

The temporary access road had been pushed through the year before and had opened up a broad strip on either side to intensive activity. The usable wood was cut and trucked out. The remaining trees, many of which were over 100 feet tall, were pulled down with chains dragged between bulldozers, pushed into windrows, and burned. The land was cleared into a regular grid of 20 hectare fields, each one surrounded by a windbreak. The trash was worked into the soil with heavy disk plows and the land was planted with rice and maize. 1976 was the first year that a full-scale attempt was made to make the area produce, and over 9,000 hectares were planted.

It was early November, and the rice harvest had been underway for a month. The earlier stages of the crop cycle had progressed with impressive ease. The seed had been sown with tractors and from airplanes. Fertilizer, and insecticides and fungicides to control pests and diseases, had been applied by air on a regular schedule. The cost had been high, but the officials had been ebullient that so large an area was being brought under cultivation for the first time. If the zone could be made to produce, all their work would be justified and the details of costs and yields could be worked out on the basis of a real, positive experience.

The past few weeks had seen one frustration on top of another. Rice Blast, a fungus disease especially serious in upland rice, had damaged a significant portion of the crop. It had rained almost every day. The combine harvesters, designed for quite different conditions in the United States, were having difficulty and the program was behind schedule.

We pulled in behind two of the harvesters that were off on the shoulder. One was seriously disabled and partially dismantled. The mechanic from the subsidiary of the bank that operates all of the machinery in the zone was up inside the maw of the other one, while five or six men looked on. We piled out to see what was the matter. The combine had pulled a stick of wood in with the rice, which had been caught in the delicate machinery inside. This was a common problem. Rice is usually harvested by machine on leveled land under irrigation. The water is drained off before the harvest and the straw dries out. Gangs of combines can work down the fields with little interference. Here in Uxpanapa the rice plants had put out large volumes of succulent vegetative growth under the influence of the applied nitrogen and that released by the recent clearing. The continuous rain had soaked the soil and the plants. The crop had grown very tall, and much of it had blown down. The combines worked their way through this mass of vegetation slowly and with difficulty. They separated the grain with a system of fans. If it was too wet, it did not separate and the rice was blown out the back with the trash. A stick left over from the recent clearing would shut a machine down for half a day or more.

The officials encouraged the workers to get moving again as soon as possible, and we got back into the truck. They were furious with the delay and angry with the machinery company. Of a total of 32 combines, only an average of 14 were in condition to work on any given day. The operators had received no special training and didn't know how to evaluate a strange noise or other problem before the damage became serious.

We turned off onto a smaller track which led into a field where work was going on. Five machines were operating, three more were stopped for one reason or another, and nearly fifty Chinantec Indians were scattered around, watching. We got out and walked over. The agronomist called the group together and began to speak. He told them that everyone had expected that the harvest would go smoothly, that there would be enough machines, that the weather would clear up. This was an especially rainy year throughout Veracruz. This land was being cultivated for the first time, and everyone was working hard to to make sure that it would be a success. The Chinantecs, as equal members of the newly formed collective _ejido_, or cooperative farm, must take responsibility. The government was doing everything that it could. It had opened up the area, built the roads, cleared the land, bought the seed, fertilizer, and other inputs. It had brought in the machines, but they couldn't do everything. Now the people themselves must take responsibility and organize themselves to harvest as much grain as possible by hand. In Oaxaca everyone had always worked by hand, on a maximum of two or three hectares. Here each family had 20 hectares.

They had the opportunity to earn much more money than before and the responsibility to provide basic food grain for Mexico, which needed it so badly. One person obviously could not harvest 20 hectares of rice alone so other people were being brought in to help, other Chinantecs who would themselves be moving into Uxpanapa in the next few years. The agronomist realized that the fallen grain was difficult to cut with machetes, so he had brought sickles, which would make the work go faster. The Chinatecs would thresh by beating handfuls of the laden panicles against oil drums, letting the grain fall in. The plants had produced well, there were at least four tons per hectare in the field; it must be gathered.

The president of the ejido stepped forward. Like the others, he was dressed in cheap cotton pants and an open white shirt over a T-shirt. He spoke quietly and deliberately. He said that both the members of the ejido and the workers who had come in from outside were very dissatisified with the situation. All of the combines together were only cutting about six tons per day. On the few days the sun had shone for a few hours and the machines had been working, there had not been enough trucks to unload into, and more time had been lost. Hand cutting was intolerably slow. In their own land they planted rice in compact clumps in rows. At harvest time, the men could work rapidly down the rows in an easy rhythm, cutting the panicles off plants and beating off the grain. Here the rice had been thickly broadcast and much of it had lodged. It was slow work to gather up the wet straw, cut off the heads, and thresh. They were paid 40 centavos (about ten American cents) per kilo harvested. No matter how hard they worked, they could not make the legal minimum wage. They were a long way from home, living in half-completed houses and improvised dormitories, and the cost of food was very high. They wanted to be paid a flat daily rate. The costs had been high through the whole season, and nobody could see how this harvest was going to yield a profit for them. They had been promised that the new land would produce large profits under mechanization and that they would live better.

The Bank manager stood up on the back of the truck. He had been working 16-hour days for months, organizing the flow of inputs, machinery, and advance payments. He saw the president of the ejido, one of 27 in the zone operating with credit, almost every day. He reminded them that the advance they received for their work was not a wage, it was part of the credit. When the time came to render accounts, all of this would be deducted from the income from the harvest. The crop was insured, the costs of production were covered so that they would not fall into debt, but they were not guaranteed a profit. Everybody would have to work together to realize the potential of Uxpanapa. Of course there would be problems the first year. Of course the people had the right to expect enough to eat and something to send home to their families. He would raise the rate to 60 centavos per kilo. The sickles would speed up the work. He encouraged them to try hard and get as much rice up as possible before it began to sprout. The weather would improve, more machines would come, more trucks were on the way; things would get better.

The officials walked around and chatted for a few moments, and demonstrated the use of the sickles. Then we got back into the truck to drive the twenty-five miles back to La Laguna, the headquarters camp. All three men were very frustrated. They could not understand why the Indians wouldn't take responsibility, why everything had to be done for them. They had been treated with every consideration since they had first been brought to see Uxpanapa by boat and helicopter three years before. The government had invested millions of pesos in roads, infrastructure, clearing, and credit, so that a group who had lived in poverty could become members of a prosperous community. They could not expect a handful of technicians to do everything, and they wouldn't even apply themselves to harvest their own rice on their own land.

We dropped the bank manager and the insurance representative off at their office, a three-room, concrete-block structure on the edge of the muddy road. Five inoperable combines were lined up in front, and two mechanics were working on a crop-spraying helicopter in the yard behind. About 20 campesinos (the Spanish word campesino for rural people corresponds roughly to the English word "peasant" but carries no value connotations) were sitting on benches outside or leaning on the parked cars, waiting for the manager to approve some paper in the complex credit system. We went on to the agriculture office, which was equally busy.

In the early afternoon the agronomist, who was also charged with the formal organization of the ejidos, invited me to attend a meeting in one of the new towns. Eight of a projected 12 of these settlements were in various stages of construction at the time. Work was suspended during the year 1976 due to budgetary problems. Each one would combine the members of several ejidos, and would provide homes for approximately three hundred families. The houses were being built by the settlers themselves with a 20,000 peso, or 1600 U.S. dollar, government subsidy. (The Mexican peso was fixed at an exchange rate of 12.50 per dollar during the planning stage of the project. In September 1976, the currency was allowed to float and devalued by nearly one half. It had stabilized at 22.80 pesos per dollar by 1978. Unless otherwise noted, the dollar equivalents used throughout this study are calculated at the earlier rate. The symbol ($) is used for the peso in Mexico.) Water, sewage and electricity were to be provided to each home. A civic center would consist of an administration building, a meeting room, a school, a health clinic, and a post office. The town we visited was one of the most isolated, about 20 miles up a side road.

The first impression was startling, of order being imposed upon nature. The jungle had been pushed back in a ragged line, exposing red soil and white limestone rock. The area had been divided up into neat blocks, outlined by the concrete poles and wires of the Federal Electric

Commission. Everything was in various stages of construction. The sidewalks had been completed before the drainage system was complete or the streets were paved, so perfectly rectangular, deep, muddy puddles had formed.

Although the settlers were free to modify any one of four basic house designs in any way they wished, they were nearly identical. A concrete floor and six reinforced concrete posts formed the basic frame. The walls were built out of irregular, locally molded concrete blocks. In many cases, the upper third of the walls were made of rough cut boards, set vertically. The roofs were corrugated asbestos sheets. They were well spaced, usually four to a block. The areas behind, which were planned for garden plots, were full of weeds, construction materials, fallen trees, and, in a few cases, poor stands or corn, sesame, and other crops. All of the buildings in the civic center were begun at once; the brick walls stood at odd angles at different heights, and nothing had yet been roofed.

The meeting was to be held in a temporary gallery for the storage of cement and other supplies. There were about fifteen people standing around when we arrived. The agronomist greeted the president of the ejido, who sent two teenage boys to get a table and some chairs from a nearby house. There were 112 legal members of this ejido, so 57 (50 percent plus one) needed to be present if any legal business was to be carried out. The president spoke rapidly in Chinantec, and about half of the workers went out to find other people. They wrapped their heads in brightly colored plastic sheets to protect themselves from the light rain.

When the simple unfinished table arrived, the agronomist and president sat down and chatted about the crops while they waited. The rice planting had been seriously affected by blast. The agronomist assured him that it would probably recover. The maize was developing poorly and unevenly. This, the agronomist said, was really their fault for not having followed his instructions about fertilization. It is a common mistake of the campesinos to wait until the plants are about eight inches tall before applying nitrogen. In this climate, it is important to put in the fertilizer at the time of planting to give the new shoots a good start. The president also complained that the truck which had been assigned to take the workers out to the fields often arrived late, and they wasted many hours.

After about 25 minutes, 51 people had gathered, including three widows who had inherited ejidal rights. All of the others were away in the old ejido in Ojitlan, Oaxaca. This was a constant problem. The presidential decree by which the area behind the dam had been expropriated specifically gave the campesinos the right to continue to cultivate their lands until they were actually flooded. The construction of the dam had been delayed, and the people were constantly moving back and forth between the two areas. Many of them regarded Uxpanapa as a somewhat irregular source of work and maintained their homes and families where they had always been. This made the daily organization of the labor force very uncertain, and greatly complicated the constant paperwork that was necessary to meet the various legal requirements of the process. The bank had been forced to send members of their inadequate staff to the old settlements, many of which are as much as a five-hour walk from the nearest road, to get the necessary signatures for the credit contracts.

The agronomist looked through the pile of papers he had brought, and asked when they thought he would be able to get a quorum together. The president talked with the other members of the leadership, and they decided that it should be possible on the following Thursday. The official said that he would be glad to come back, but he had many obligations all over the zone and he would take the opportunity to explain the purpose of the meeting to save time the next week. He had been personally instructed by Ing. Jorge Tamayo, the director of the Papaloapan Commission, which was in charge of the dam and resettlement, to legally organize the collective ejidos of Uxpanapa into a regional association, or Union of Ejidos. As they all knew, each family was granted rights to 20 hectares of land in the new ejidos, but it would never be divided into individual parcels. Following the framework set down in the new Agrarian Reform Law, all of the resources would be managed cooperatively as a single productive unit.

In comparison with the great majority of campesinos in Mexico, they were uniquely privileged. In compensation for the flooding of their homeland, the Chinantec people would receive a total of 28 million pesos. The value of their personal immovable property was to be paid in cash. Many of them had already received this money; the documentation for the balance was being processed in Mexico City and would come through in the course of the next year. The assessed value of their land and the communal property of the old ejidos was put into a common fund. This would be augmented by their share of the income from the labor which was being cut and taken out.

The Union of Ejidos would receive this money and invest it in an agro-industrial center, with the technical assistance of the appropriate government agencies. Anything that is good business should come under the control of the people. They needed a fleet of buses to provide transportation; it should belong to the campesinos. They needed tractors to work the land; they should belong to the campesinos. Why should they sell unhusked field rice at three pesos per kilo, when polished rice sold for $7.50? A rice mill should be built, and it should belong to the campesinos. Already an outsider was finding it profitable to run a gas station; the campesinos should build their own and keep these profits for themselves. The Union would be organized democratically, just like a larger, richer, and more powerful ejido. By uniting they would become one of the most powerful local organizations in the whole country. They would be able to secure large loans, and to exert political and economic power far beyond the capacity of an individual ejido, not to mention an individual person. Important officials from the capital would be coming to Uxpanapa in a few weeks to legally establish the union. The majority of the members of each ejido must approve the transfer of funds, and elect the representatives.

Obviously somewhat awed by the forcefulness of this speech, the ejidatarios sat quietly for a few moments. Then a man of about 40 got up and began to speak very slowly, his Spanish accented with the short, breathy sounds of his own language. He said that many of the people didn't much like the collective system. In Ojitlan each family had its own parcel, and took care of its own affairs. Here, some people were lazy and took advantage of those who worked hard. Also, if the ejido had all this money, why couldn't they buy some of the things they so urgently needed? The round trip back to his old home, where his family still lived, cost him about three days' pay. With all the rain, they often didn't get that much work in a week. They often had to wait several days for the bank's tractors to arrive. Why couldn't they buy a truck for transportation, and some tractors to work?

The agronomist spoke up again. He said work responsibility was the greatest, and really the only, problem of cooperative organization. The campesinos of Mexico had gotten into bad habits. Factory workers must get to work on time. If they are late one day, they get warned. If they are late again, they get their pay docked. If they are late a third time, out they go. In the country, people have always taken things too slowly. They get to work, look around, work a little, take a rest in the shade, and go home when they feel like it. Mexico would never progress like this. They must assume the discipline of responsibility, and really put their backs to the wheel.

As to the tractors and trucks, these should belong to the Union. Suppose some fertilizer is needed. Should some merchant be allowed to make money delivering it to our door? No, we should go to the factory in Minatitlan and buy it at the wholesale price. Should we send ten trucks down there and have each one pick up four or five sacks? No, the Union should take care of this. It should take orders from all the ejidos and send down a truck to buy it all at once, perhaps in bulk which would be much cheaper. Should ten trucks drive up to Ojitlan each weekend? Can we really afford such luxury? No, the union should buy a bus and organize a regular route. None of these services should be free; the Union should be run like a business. The union must not go through its capital and have nothing to show in five years but some broken-down machinery and an empty bank account. Its operations should be run for profit, and the profits should belong to all of you. I have many things to attend to now. Discuss it among yourselves, talk to the ejidatarios who could not make this meeting, and be ready by next Thursday to sign the papers. You are free not to join, but I am sure that if you think about it, you will.

We shook hands with everyone and got back into the truck. On the way back to the office, the agronomist told me that the process was bound to be slow and somewhat difficult, but that he was sure that it would work out well. He said that these Indians were always suspicious of outsiders and new ideas. He did not expect all the ejidos to join immediately, but as soon as they saw the real benefits received by those who did, they would all come in. It was important that all the capital be kept together rather than being distributed among the ejidos, where it could easily be frittered away.

In the late afternoon, I went out by myself to interview the President of one of the ejidos which had been settled about ten years before, when it was almost entirely virgin wilderness. It was a very different kind of town from the one I had just visited. Its houses were built of bamboo with very steeply pitched thatched roofs. They were laid out in an irregular line on a bluff overlooking a river. A new wooden school was set up on a hill behind. Tropical towns are never neat; the paths are irregular and muddy, pigs run freely, and at first it is difficult to recognize the gardens as such. On closer look, the tangled growth reveals a great variety of vegetables, fruits and herbs. After some asking around, I located the president, whom I had met several days before at the bank. He was a short man with closely cropped black hair and calm, intelligent eyes. He seemed pleased that I had come, shooed away a crowd of curious children, and asked his wife to bring some chairs outside. Within a few moments, three other men stopped by, one by one shook hands, and sat down. They were obviously close friends, and joked and chatted among themselves.

I asked them how they had come together to settle in such an isolated area. The president explained that most of the ejidatarios were Totanac Indians from the northern part of Veracruz. They had worked as day-laborers on farms and found occasional work in the town, loading trucks. Many years before, they had formed a group and applied to the agrarian authorities for land. The tenancy structure in their home area was very complex, and two separate petitions for an ejido had been endlessly tied up in litigation and bureaucratic paperwork. In the late 1960s, they were offered the chance to move into the Uxpanapa area, which had recently been divided into ejidos. Six of them had taken the train to Jesus Carranza, a small port on the Coatzacoalcos River. From there, an official accompanied them in a launch as they spent a week exploring five possible sites in what was then an unbroken forest. They had chosen this area because of its location on a bluff overlooking the river near a broad band of good, alluvial soil. They returned home to make the arrangements and moved down the following season.

I asked how many people had been part of the original group. The men consulted among themselves for a few moments before agreeing that 42 families had originally signed the papers. There was much confusion at first. The government provided legal rights to the land and the cost of transportation, but very little else. In the first years, each family cleared between four and six hectares of land by hand. This was about as much as they could use, because there was virtually no market for their products, due to the high cost of transportation. Merchants would occasionally come up the river in launches to barter oil, salt, kerosene, medicines, and other necessities for corn at 40 or 50 centavos per kilo, less than a quarter of the official support price. The people also gathered barbasco, a root found in the forest, which is used for the production of birth-control pills. They supplemented their diet with game. The first years were very hard. There were no schools, no doctors, and the river was impassable for two or three months of the year during the dry season. Many members of the original group had not been able to accept the hardships, and left after a few months. Other relatives and friends had replaced them; there had been a lot of coming and going.

I asked them how the arrival of the Papaloapan Commission had affected their lives. They looked at each other and seemed a little unsure how to answer. I said that, well, you have a road now. The President said yes, that this was very important. Merchants now came in trucks to buy their harvests at much better prices. When somebody got sick they could get medicines; their lives were much more normal. One of the other men said that the road was still over a kilometer from the town. He added that when the government had come to take out the timber prior to clearing the land, they had promised them 600,000 pesos and permanent gravelled access roads. In fact, they had never even been told how much the wood had sold for. The commission was pressuring them to release the money to the Union of Ejidos for the benefit of the zone as a whole, but they wanted it to meet their own needs. They had been able to buy a truck and a tractor with some of these funds, but the roads were muddy tracks which had never been properly graded.

Little by little, the conversation began to pick up, and we spent several hours discussing their experiences and the issues important to them. The president did most of the talking, but he would consult with the others and they would often spend some time arriving at an exact detail or figure. When the commission came, the colonists were offered all of the same benefits being given to the Chinantecs: housing, electricity, medical care, schools, credit, and so forth. In return, they had to agree to become part of the integrated, centralized structure that was being set up. They had worked and suffered for the right to farm their own land in their own way, and a variety of conflicts with the government had arisen over the organization of their settlement, and more significantly, over the type of agriculture and kind of credit arrangements that would really meet their needs.

As soon as the project began, the commission sent surveyors out to draw lines as the basis for their clearing and resettlement maps. In this ejido, as in several other cases, the results conflicted seriously with the documentation they had originally received. Areas that some members had laboriously cleared by hand were excluded from their property, and the Commission proceeded to clear with machines and plant crops on land the ejidatarios believed was legally theirs. At the same time, a census was taken of the existing population. According to the settlers, members of the community who happened to be away the day the helicopter came were arbitrarily excluded. Each family was legally entitled to 20 hectares, and several Chinantec families were settled into the vacancies. The original group would have preferred that these places be left open for their own children, and to allow for land that was too steep or rocky to plant.

The commission had set up guard stations, manned by a special police force and a small contingent of the army, to protect the newly opened area from invasion by landless campesinos before the resettlement program was completed. Identity cards had been issued; it was complicated to get permission for friends and relatives to visit. One of the men wryly observed that it was easier to cross into the United States than into Uxpanapa.

The colonists were being asked to move into one of the new towns, which was being constructed about six kilometers away, to receive services such as housing, running water, and electricity. They were unwilling to go; they liked their location on the river. They had invested tremendous amounts of time and energy in clearing the land and planting their gardens, and did not want to live among strangers with a different language and different customs. The president commented that the Chinantecs were dominated by their own traditional leaders or caciques, and that the unity that had kept the original group together in the early years was breaking down into petty conflicts. He was negotiating with the commission to drop the 20,000 peso housing subsidy in return for electricity in their present location, but they were not sure if this idea would be accepted. One of the men said flatly that he would take his family out to live in the fields rather than move into the new town.

As farmers, the colonists were very frustrated by their problems with the agricultural program of the commission and the credit policies of the official bank. In their first years in the area, they applied their experience from their home area to the problem of growing enough food. Each family cleared several scattered small plots. This allowed them to take advantage of different soil and drainage conditions to plant different varieties of maize, as well as rice and other crops, at different times of the year. Without the resources to use purchased inputs such as fertilizer, or any source of power except their own labor, they were able to survive.

The commission cleared 450 hectares in the ejido with heavy machinery, saving many years of heavy labor, but in three years the people had made no profit in these areas. Credit was tied to a whole series of complex technical and bureaucratic procedures set for the region, which had not been adapted to the specific natural conditions or the needs of the settlers. New varieties and a highly mechanized system had not worked well. Costs had been very high, and the yields were lower than in the plots cultivated using traditional technology. Although the credit was received through a collective credit contract with the ejido as a whole, and the fields were prepared in large units with the bank's tractors, the leadership informally divided the land and the materials among the members. Many individuals were beginning to feel that they would do better entirely on their own.

After a couple of hours of discussion, the president invited me out to look at the fields. We drove for about three kilometers along the main road, and then turned off into a track that led to a large field of maize. He was particularly anxious to show me the difference between the hybrid and native varieties. The hybrid stand was extraordinarily uneven - the seed had not germinated at all in large patches - and many of the plants were small, yellow, and sickly. The leaf covering of the ears was not complete at the tip, where moisture and insect larvae enter and damage the grain. The native varieties were much more resistant to a variety of problems. They did best on soils cleared by hand rather than by machine, but consistently out-yielded the hybrids. We chatted for a few moments about the crops, and I gave him a ride back to the headquarters camp, where he had some business with the bank.

Uxpanapa is one of several large, integrated projects the Mexican government has undertaken in the rainy tropics in recent years. The construction of the Cerro de Oro dam and the need for resettlement is the specific reason to open up this area, but a growing national food deficit and rapid population growth are putting increasing pressure on the land and other resources in the Southeastern region of the country. The Mexican tropics are relatively small, and do not present the large-scale challenges of the Amazon basin and other areas of the world. Eighteen million hectares form less than ten percent of the country, but include between eight and ten million hectares of sparsely populated, potentially productive crop and pasture land, and enormous water and forest resources. In theory, a program of this kind allows the government to concentrate its investments, increase agricultural production through mechanization and supervised credit, provide basic services to the population, and build the basic infrastructure for regional growth.

This plan has generated controversy in political and academic circles in Mexico. The Papaloapan Commission has been accused of systematically destroying an Indian culture, of destroying one of the last areas of tropical rain forest in the country, and of devoting a disproportionate quantity of public funds to a project of limited impact and uncertain success based on inappropriate technology imported from other regions.

As of this writing (1978), the Uxpanapa project has been underway for less than five years. Within this short perspective, it is not possible to make a formal evaluation of the costs, benefits, and distribution of benefits. Nevertheless, important and irreversable decisions have been made. A certain style of development - a certain methodology - has evolved out of the priorities of the Mexican state, the growing crisis in the agricultural sector as a whole, previous experience in tropical colonization, and the particular institutions involved.

This study will explore the background of the government's policy and three very different modes of tropical agriculture. The first is the traditional system of the Chinantecs in their homeland, the second is the system used by the spontaneous colonists in Uxpanapa to survive in the wilderness, and the third is the mechanized, collective system of the project itself. Uxpanapa is a useful framework within which to examine the issues involved in introducing modern technology into the rainy tropics. The experience may help to define the most appropriate strategy for the development of these regions, both in Mexico and in other parts of the world.

2 Mexico's Agricultural Crisis

The pressures on the Mexican government to increase production in the tropics are intensifying. On the national level, the total production of basic foods has not been able to keep pace with the growth of population and total demand, and imports have been increasing steadily. The level of living of the majority of the rural population is very low, promoting a massive migration to the cities, where industry is concentrated but where insufficient work is available. In 1978, it was estimated that only about half of the work force in Mexico City was fully employed (Huey, 1978). The problems of the agricultural sector have reached crisis proportions in the past ten years, after three decades of remarkable growth and 50 years after an agrarian revolution which laid the basis for one of the most comprehensive land reform programs in the world.

Beginning in the 1940s, development policy shifted away from a broad strategy based on a viable peasant agriculture, to the concentration of capital and modern agricultural technology on a small fraction of the total farm land, favoring large, privately owned units in regions suitable for irrigation. This approach more than adequately served the interests of an elite whose considerable fortunes were tied to the rapid industrialization of Mexico. Between 1940 and 1965, agricultural production increased at an average rate of over 5 percent per year while GNP rose by over 6 percent, figures well above the 3.1 percent rate of population growth and among the very highest in the world at that time.

This performance was made possible in part by massive public investments in infrastructure, and in part by the development of efficient technological packages which include high-yielding varieties of wheat and other crops, mechanization, fertilizers, pest control, and other modern inputs. The "Green Revolution" provided food to the urban areas at relatively low cost during a period of rapid industrial expansion. Nevertheless, it both masked and contributed to increasingly serious distortions in the allocation of resources, both between regions and between sectors of the rural population. The growth rate dropped off rapidly after 1965; in both 1976 and 1977, agricultural output declined by about one percent per year. Mexico is importing an increasing proportion of the basic food grains needed to feed its people. These trends are illustrated in figure 2.1. Government policy has been actively seeking out alternatives, including a greater attention to the peasants who worked very close to or below the subsistence level - who constitute the overwhelming majority of all farmers - and to under-exploited regions such as the tropical Southeast.

GEOGRAPHY: HIGH AND DRY

Mexico is dominated by the great chain of mountains that runs, with a few interruptions, from Alaska to the tip of South America. The country is divided into sharply contrasted zones: hot coastal plains, sharply rising escarpments, and interior plateaux. The land is further broken down by the irregular topography into hundreds of ecologically distinct sub-zones (fig. 2.2). Two-thirds of the total area is on slopes greater than ten percent; over one-quarter is on agriculturally useless slopes greater than 25 percent. High altitudes produce a temperate climate in much of the country, but over three-quarters of the potentially arable land is arid and only a few zones are suitable for irrigation. Out of a total area about one quarter that of the United States, about 15 percent is potentially arable and about 9 percent is actually cropped. This amounts to 17 million hectares - a little bit more than that cultivated in the state of Iowa - on which to support a rapidly growing population of 65 million (McBride, 1923; Tamayo, 1964, p. 28; Wellhausen, 1976, p. 130).

There is no natural breadbasket - no large, fertile area appropriate for the intensive rain-fed cultivation of grain crops analogous to the Midwest of the United States or to the plains of central Europe. Since before the Spanish conquest, the Mexican people have been concentrated in the dry, temperate highland basins around the cities in the center of the country. Over half of the population and over half of the cropped land are found in this zone, which receives less than ten percent of the total rainfall (Wellhausen, 1976, p. 139). Generation after generation of subdivision has resulted in very small farms averaging between one and five hectares, which are cultivated using traditional methods based on hand and animal labor.

The southern highlands, where the Indian people are concentrated, are the poorest areas in Mexico. The rainy tropical lowland strips along both southern coasts were disease-ridden, sparcely populated, and almost entirely undeveloped until a generation ago. Recent projects have not yet established intensive production in these areas; they are dedicated primarily to extensive and inefficient cattle ranches, sugar cane plantations, and primitive subsistence agriculture based on slash-

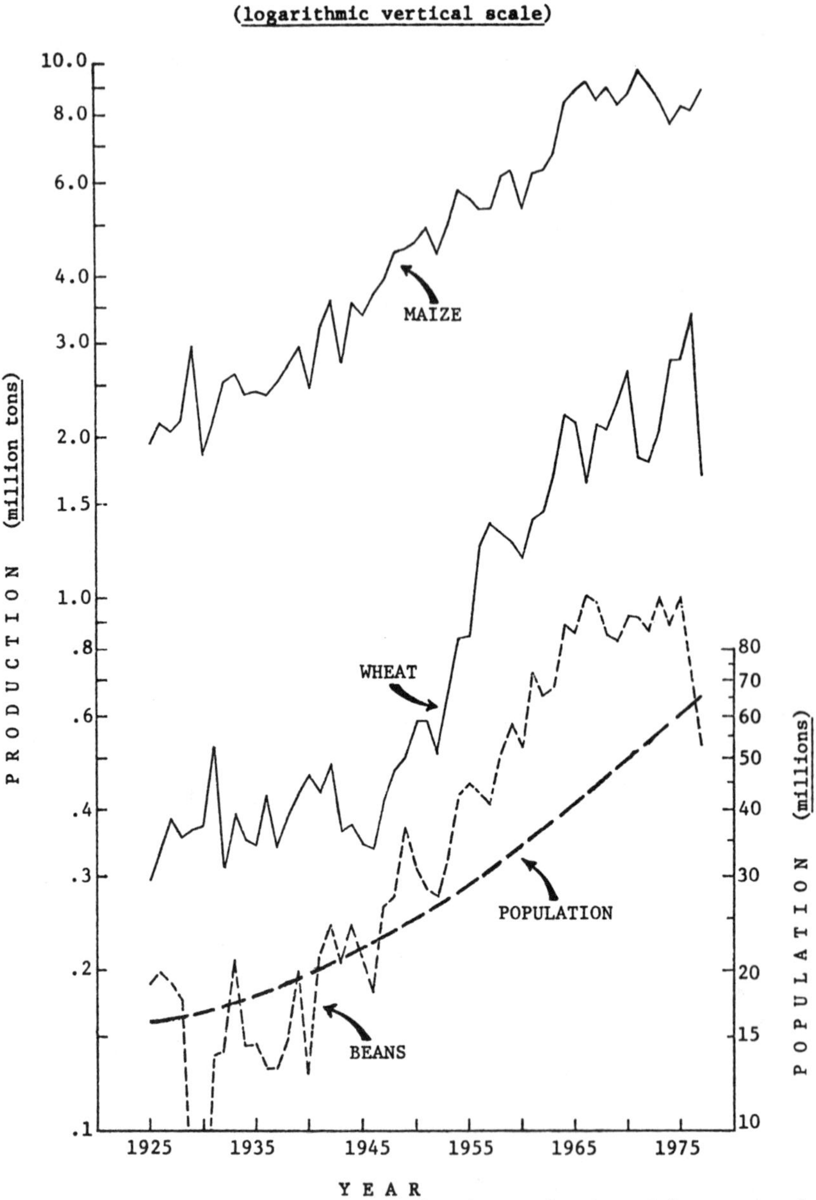

Fig. 2.1. Mexico: production of maize, wheat and beans, and growth of population, 1925-1977.

Source: Thomas T. Poleman, "Food, Population, and Employment: Some Implications for Mexico's Development" (Ithaca, N.Y.: Cornell Agricultural Economics Staff Paper No. 78-19, 1978), p. 8.

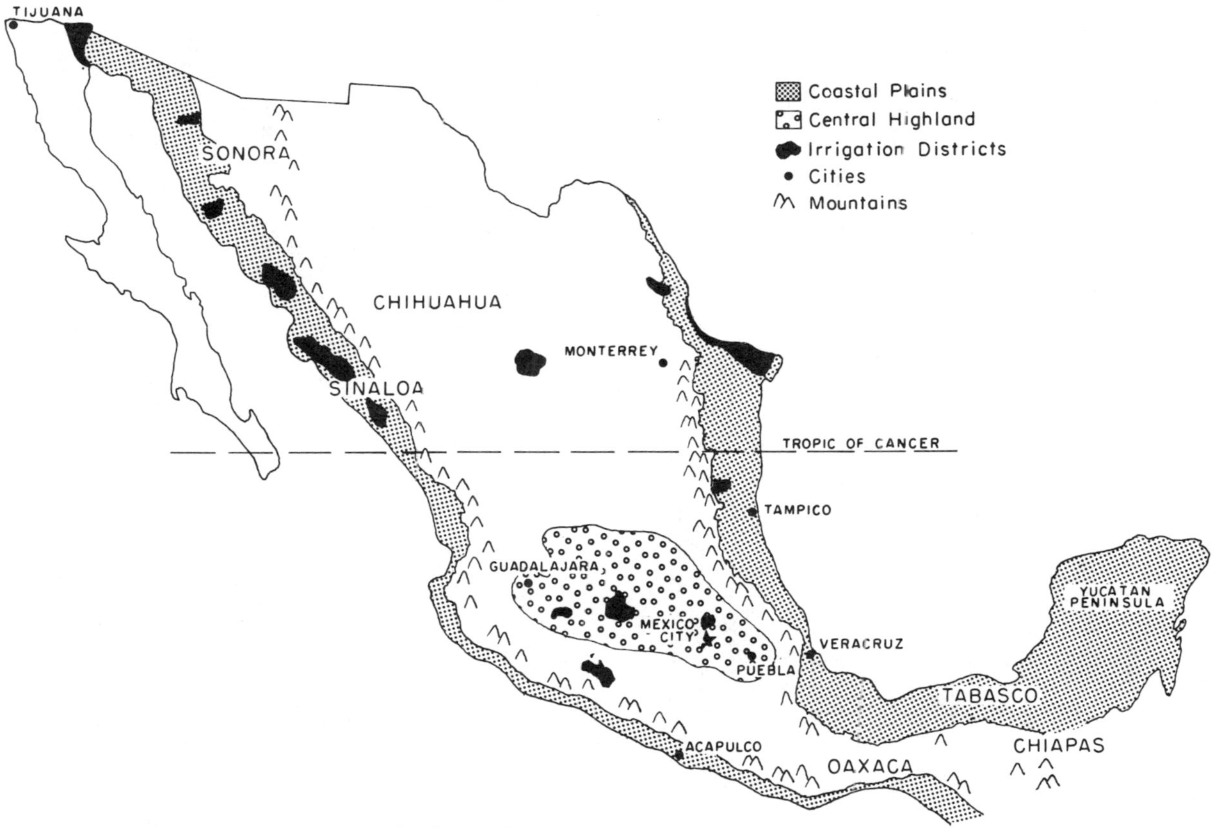

Fig. 2.2. Mexico: major geographical regions.

and-burn techniques. The porous limestone soils of the Yucatan peninsula, which was long isolated from the rest of the country, have little potential for intensive agriculture apart from the production of the traditional henequen and sisal fibers, which face an uncertain world market.

The Importance of Irrigation

Irrigation has been the key to highly productive agriculture since pre-Colombian times. The most remarkable increases in output in the post-war period were concentrated in the north, particularly in the states of Sinaloa and Sonora, where the government invested heavily in building dams on the short rivers to water the arid valleys along the west coast. In the years between 1940 and 1970, from 10 to 20 percent of the federal budget was directed to the agricultural sector. Between 80 and 95 percent of this was invested in the physical construction of irrigation systems (Wilkie, 1967, pp. 131-136). Mechanization, credit, research, extension, and other services have been concentrated in these oases. An intensive program of research sponsored by the Rockefeller Foundation developed varieties of wheat capable of very high yields when grown using a carefully controlled package of water application, fertilization, and other practices. The yields of other crops have increased significantly.

Capital-intensive, technically sophisticated irrigated agriculture in federally-administered districts generates more than a third of the value of the commercial farm output in the country, although it occupies less than ten percent of the total cropland (Fernandez y Fernandez, 1978; Wellhausen, 1976, p. 139). (See fig. 2.3). The modern sector, which only accounts for about seven percent of all Mexican farms, is concentrated in the irrigation districts in the north (Cisneros, et al., 1970, chart 3-1). These farmers specialize in relatively high-value crops, most of which are directed to a small, high-income domestic market or are exported. There has been a dramatic trend away from basic food crops in the irrigation districts in the past 15 years (Silos, et al., 1975, chart 21). The best dam sites have already been developed and the cost of bringing new land under irrigation has risen to over 3,000 dollars per hectare (Tamayo, 1976b). Between 1970 and 1976, the government developed 600,000 hectares in smaller, lower-cost projects, but their impact was neutralized by organizational problems and the abandonment of marginal land in rain-fed areas (Banco Nacional de Mexico, 1977). Rapid increases in production through irrigation can no longer be relied upon to meet the nation's growing need for food in the years to come.

AGRARIAN REFORM

Land Reform has been one of the principal elements of the agricultural policy of every presidential administration since the Revolution. Over 90 million hectares had been granted to nearly three million beneficiaries by 1976. Nevertheless, as figure 2.4 illustrates, the basic productive resources are very poorly distributed. More than 85 percent of the farmers in the country are unable to support their families on their small plots using the technology that is available to them (Reyes, et al., 1974, p. 953). A study made by the Bank of Mexico in 1968 concluded that 52 percent of the landholders made less than 3,500 pesos, or 380 dollars per year, from the sale of crops. Although small farmers produce some of their own food, the average family in this group sold over half of their production. Ninety-five percent of the nation's

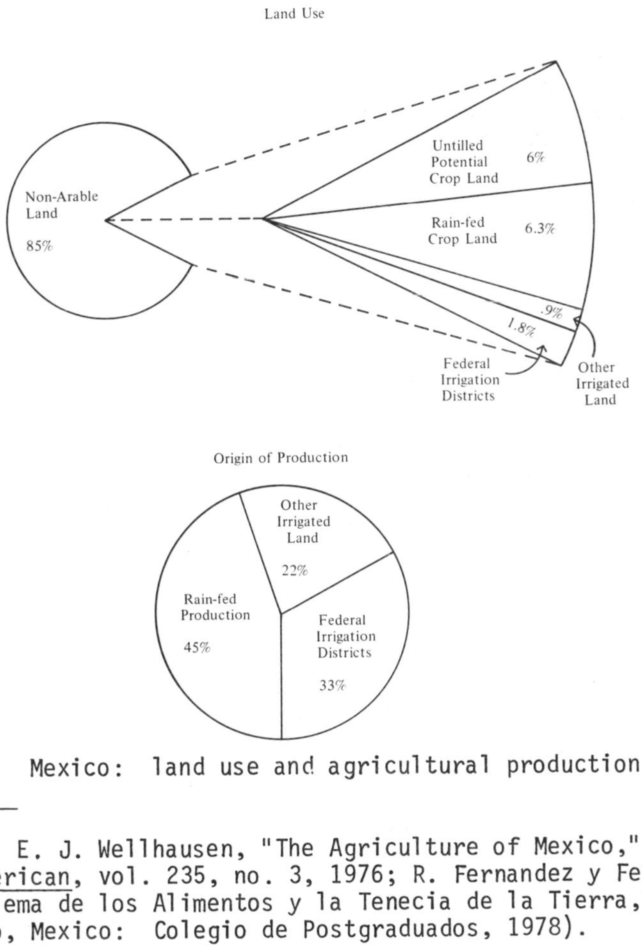

Fig. 2.3. Mexico: land use and agricultural production, 1976

Sources: E. J. Wellhausen, "The Agriculture of Mexico," Scientific American, vol. 235, no. 3, 1976; R. Fernandez y Fernandez, "El Problema de los Alimentos y la Tenecia de la Tierra," (Chapingo, Mexico: Colegio de Postgraduados, 1978).

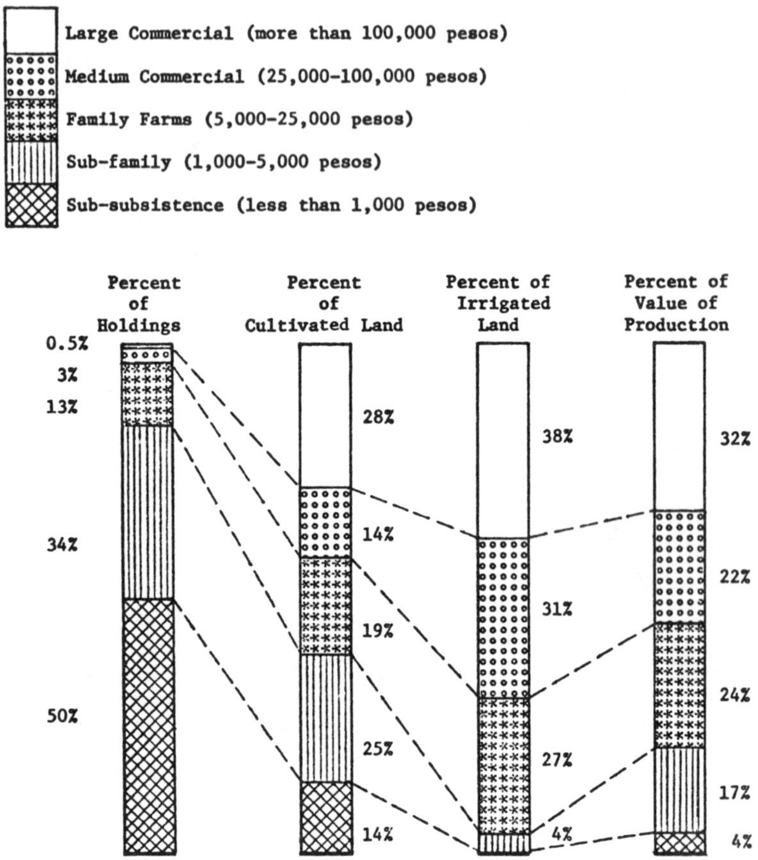

Fig. 2.4. Mexico: distribution of cultivated and irrigated land and value of output by type of holding, 1960.

Source: Thomas T. Poleman, "Food, Population, and Employment: Some Implications for Mexico's Development" (Ithaca, N.Y.: Cornell Agricultural Economics Staff Paper No. 78-19, 1978), p. 9.

farmers made less than 9,300 pesos, or $740 per year (Rodriguez Cisneros et al., 1970, Chart III-1). The agricultural sector is polarized into extremes in many variations of rich vs. poor: irrigated vs. rain-fed production, north vs. south, large vs. small, modern vs. traditional technology, and private vs. community land ownership. Capitalist development inevitably concentrates resources in its early stages, but in few countries have the benefits of growth been so poorly distributed. How is this possible under a stable democratic government formed after a successful revolution in which two million campesinos died under the banner of "land and liberty"?

The Land-Holding Villages

Over 90 percent of the farmers in the country are in the traditional sector, which has been on the margins of the national economy since the Spanish conquest. The invaders opened up new areas to meet the needs of their own food economy, but they left the native communities to feed themselves much as they had always done on the basis of communal land ownership. The people became racially mixed, but agriculture continued to revolve around the village (Whetten, 1948, Chapter IV). A great variety of systems based on hand and animal power developed over many generations of experience to take advantage of the local ecological conditions and the available resources. Both the Spanish landowners and the Church were content with a share of the product, and did not try to change the basic structure of these communities.

In the 1860s, President Benito Juarez promulgated the Reform Laws, which tried to break away from the semi-feudal, locally-oriented structure of rural Mexico by requiring the individual ownership of land. He hoped that this would release entrepreneurial spirit in the new family farmers and form the basis for progress on the model of Europe and the United States. The abrogation of the ancient rights, instead of strengthening the position of the poor, provided an opportunity for the elite favored by his successor, Porfirio Diaz (1876-1911), to buy or despoil the campesinos of their land and enlarge their estates, or _haciendas_, to enormous proportions in many parts of the country. At the same time, the construction of the railroads, the improvement of communications, and the effects of industrialization on the need for exports created conditions that favored the development of capitalist enterprises in agriculture. Large areas were brought under irrigation. The intensive cultivation of sugar cane, cotton, and other commercial crops required a large landless labor force. By 1910, over 90 percent of the rural population in what was still a predominantly agricultural country owned no land at all.

The Revolution and the Ejidos

The Revolution was fought primarily by the campesinos, who wanted to destroy the power of the exploitative landlords and get their land back. Zapata and the other agrarian leaders formulated increasingly specific programs for rural development. Although their promise was never fulfilled, they did determine the institutional structure of the post-revolutionary society.

The most basic and universal demand - the restitution of the lands taken from the rural communities in an inalienable form of tenure - was met through the formation of ejidos. This word, which had referred to the pastures, woodlots, and other common property of a village, was now used to denote a specific form of tenure defined by the Constitution of 1917 and subsequent laws. Every rural community was given the right to petition the federal government for land within a seven-kilometer radius of the principal settlement. If the request was accepted, all males over 16 who met certain requirements formed an ejido. They were given life-time usufruct rights to the land they needed. By law, ejidal rights cannot be bought, sold or rented, but they can be passed on to one's heirs. In most areas, each family was granted an individual parcel. In theory, a legal minimum size was set for each ecological zone to insure that every family would have adequate resources. In fact, however, there has not been enough good land to go around, and most plots are too small to support the needs of a family. The members, or ejidatarios, meet formally at regular intervals to make the decisions that affect the groups, and elect a president and other officials to represent their interests.

In its simplest form, this institution does provide a democratic framework for the equitable access of the rural population to the land. A maximum size restriction was placed on private farmers, and the haciendas were broken up and redistributed. The question remained, however, as to how the newly created reformed sector was to be organized to participate in a process of agricultural development. This problem was not really faced until the late 1930s, when President Lazaro Cardenas (1936-1940) made rural development one of the cornerstones of a policy of active government participation in the economy.

Cardenas greatly accelerated the agrarian reform process, redistributing nearly three times as much land in his six-year term as his predecessors had in the previous twenty. Between the census years 1930 and 1940, the proportion of the total arable land in ejidos increased from 13 to 48 percent, and the number of landless campesinos dropped from 68 to 36 percent of the rural work force (Hewitt, 1976, p. 4; Osorio et al., 1974, p. 501). A special bank was created to provide working capital and technical assistance to the new ejidos. Irrigation systems, roads, schools, health clinics, and other public works were undertaken in rural areas at an unprecedented rate. Cardenas supported the political organization of the campesinos, founded the CNC (National Confederation of Campesinos), and made it one of the three constituent elements of the ruling party.

In response to a general strike by well-organized workers, Cardenas included a type of farm in the land reform that had previously been exempt; the large, irrigated, vertically integrated operations producing high value commercial crops such as cotton. The beneficiaries were not local villagers, but the landless laborers who had worked for wages. Collective ejidos were organized as production cooperatives to maintain and improve upon the economies of scale of the original operations. Some of the most productive agricultural areas in the country were very rapidly reorganized in this way, and a variety of institutions and organizations were set up to assist them. In 1940, for the first and only time, the reformed sector as a whole both produced more than half of the value of national agricultural output and was more productive on a per-hectare basis than the private farmers (Reyes Osorio, 1974, p. 501). The overwhelming majority of the ejidatarios in the country cultivated their newly acquired family plots using traditional technology, but first steps were being made to provide them with capital, technical assistance, and other services.

Postwar Policies

The process of redistribution of productive resources and real wealth in the countryside was reversed during World War II. Industrial production was stimulated to serve both domestic and foreign markets whose previous sources of supply had been cut off. The need to restructure the economy to take advantage of this opportunity coincided with the interests of the new urban middle class which had benefited from the elimination of the old elite during the Revolution. At the same time, it was widely believed that industrialization was more or less synonomous with development, and that private enterprise was intrinsically more productive than broadly based, cooperative ventures. President Manuel Avila Camacho (1941-1946) stated in his inaugural address that the future of Mexico would be based on "the vital energy of private initiative" and that his government would protect private agricultural properties. Agriculture was to become "the foundation of industrial greatness" (Hewitt, 1976, p. 6).

Concentrated industrialization in the cities proceeded very rapidly. The compound annual growth rate in manufacturing jumped from 4.6 percent in 1930s to 8 percent in the 1940s, and stayed above 7 percent for the next decade. Profit rates were extremely high because of the almost unlimited supply of cheap labor from rural areas, tightly controlled labor unions, and very low effective tax rates (Reynolds, 1970, p. 166). Nearly 16 percent of the rural workers left the countryside to work in the cities or in the United States during the war (Hewitt, 1976, p. 9). This rapid change produced high rates of inflation and disrupted the established patterns of food production and distribution. By the end of the war, Mexico was importing between 15 and 20 percent of its basic food grains (Poleman, 1978, p. 8). Continued industrial expansion depended on cheap food to keep wages down to maintain profits and savings, and on the conservation of foreign exchange to import machinery and raw materials. The response of the government was to concentrate resources in the development of highly productive agriculture on private farms in a limited number of irrigated oases. This strategy permitted Mexico to meet growing domestic demand and to maintain export earnings from agriculture for nearly thirty years.

Nevertheless, the redistribution of land through agrarian reform continued on a smaller scale, primarily for political reasons (table 2.1). The stability of Mexican one-party democracy depends upon the ability of the government to respond to discontent in any local area before it can become an organized challenge. The capitalist development strategy of rapid industrialization depends upon the concentration of resources, services, and income in the cities. This contradiction was met in the 1950s and 1960s by giving the rural people their most basic demand - land - in small parcels in increasingly isolated and unproductive areas. Capital and agricultural resources such as credit, tractors, and improved techniques were concentrated where the return was the greatest and the fastest (Stavenhagen, 1970, p. 250). Most basic food crops such as maize and beans continued to be produced in rain-fed areas, which were repeatedly sub-divided into tiny plots, or *minifundia*, which were too small to absorb modern technology or to support a family using traditional methods.

The agrarian reform has kept the rural people, who now constitute about 40 percent of the work force and yet account for only 10 percent of GNP, from making demands the socioeconomic structure cannot meet (Gordon, 1978, p. 30). The ejidatarios are entirely dependent upon the government. Legal titles, credit, school, roads, electricity, and other services are in short supply and can only be obtained through an endless series of petitions, visits to Mexico City, and negotiations with the tightly controlled political organizations that officially represent the campesinos. These mechanisms give the ejidatarios a stake in the system, even though it has not been able to involve them in the process of agricultural and social development that Zapata and Cardenas envisioned.

The Decline of the Collective Ejidos

With a few exceptions, the collective ejidos that were set up in the 1930s in the modern, irrigated sector have not been able to compete with the private farmers and have disintegrated. (For detailed studies of the history of these cooperatives, see Eckstein, 1966; Hewitt, 1976; Restrepo and Eckstein, 1975; Glantz, 1974; and Simpson, 1937.) The cooperative management of farm enterprises is a complex challenge, especially in a highly competitive environment, and requires discipline, able leadership, and comprehensive support. The Cardenas administration set up a series of institutions including the Ejidal Bank, financial

Table 2.1. Mexico: Redistribution of Land to Ejidos, 1915-1976

Period	Administration	Land Distributed		Ejidatarios Benefited
		Area	Percent Arable	
		(millions of hectares)		(thousands)
1915-34		7.5	-	946
1935-40	Cardenas	20.0	26	776
1941-46	Avila Camacho	6.0	19	123
1947-52	Aleman	5.4	21	109
1953-58	Ruiz Cortines	5.7	26	226
1959-64	Lopez Mateos	9.0	19	284
1965-70	Diez Ordaz	18.1	9	322
1971-76	Echeverria	16.2	-	-
TOTAL		91.0		2,786*

* Through 1970.

Source: Inocencio Higuera Ciapara, "Alimentos, Poblacion, y Empleo," (Ithaca, N.Y.: Cornell Agricultural Economics Staff Paper No. 77-1, 1977), p. 6.

cooperatives (SICAS), and Unions of Ejidos to help the collectives to combine the advantages of large-scale operations with the equitable distribution of the benefits (Gomez Tagle, 1974). The slogan of the time was "Haciendas without Bosses." In spite of a series of organizational problems, including a very high man:land ratio and the right of the old land owners to keep the best land and the core of the farm installation for themselves, production increased significantly in the first years.

In the 1940s, government policy turned very sharply against the cooperatives; organizers were even murdered by agents tacitly supported by President Miguel Aleman (Hewitt, 1976, p. 194). The institutional system was no longer committed to helping the collective ejidos and became an inflexible, inefficient bureaucracy which prevented them from competing successfully with the private sector at a time when enormous profits were being made in irrigated agriculture. This made them captive clients of the Ejidal Bank, which shared the technical management of the farms with the elected leadership. The bank bought seed, fertilizer and other inputs in bulk and delivered them to the fields. Tractors from central machinery stations did the field work on a contract basis. At the end of the season, the bank took possession of the harvest, sold it, deducted all the costs, and divided the residual among the ejidatarios in proportion to the number of days they had worked. Inefficiency and corruption led to the supply of inappropriate and expensive inputs at the wrong time, poor technical coordination, long delays in liquidation, and mounting debts.

In the 1950s and 1960s, high-yielding varieties were introduced into the irrigation districts. The new seeds required an increasingly sophisticated and expensive package of practices to achieve profitable yields. The private farmers, who enjoyed powerful political connections, organized credit unions and received government subsidies to buy machinery and other inputs at low cost. They were in good communication with the experiment stations which were constantly modifying the practices required to make the new technology profitable. Responding to exactly the same kind of spiral which has led to the enlargement of farms in developed countries such as the United States, they quickly became overcapitalized and began to buy or rent more land so they could improve the return to their machinery and irrigation equipment.

The inflexibility of the ejidos and their support structure prevented them from taking advantage of the same services. The ejidatarios were under heavy official pressure to dissolve the cooperatives and work their plots individually. Frustrated by the inefficiency of the bank, high debts, and low incomes compared to the prosperity around them, many did so. A newly independent ejidatario found himself with fifteen to twenty hectares of excellent land, a share of machinery of the old collective - a tractor without a plow, for example - and very little working capital. He would rent his parcel to a neighboring private farm or rich ejidatario. This is illegal under the Agrarian Reform Law, so he received considerably less than the full marginal value of his land to his "tenant" as rent. He probably did not work at all, and would borrow from his "tenant" to meet his consumption needs, thus committing himself to renting out his parcel again the next year.

As this process became more widespread, it pushed up the profitability of the favored sector and accelerated the decline of the ejidos. By the mid-1960s, it was estimated that 80 percent of the ejidatarios in the Yaqui Valley in Sonora had given up control of their land (Hewitt, 1976, p. 21). A similar situation is found in the other irrigation districts in the North. It is not surprising that these regions have been the focus of campesino unrest in recent years, even though the general level of living is by far the highest in rural Mexico.

THE TRADITIONAL SECTOR

The competitive pattern of the traditional farmers in rain-fed areas was unfavorable to begin with, and has deteriorated rapidly. This sector includes both ejidatarios working plots of steadily decreasing size, and private farmers, most of whom are also minifundistas with very limited resources. Government policy has kept food prices down to control the urban cost of living and to encourage industrialization. The price paid to farmers for maize declined 33 percent relative to the general price index between 1963 and 1972 (Fernandez y Fernandez, 1978, p. 6)

The small farmers have not been able to increase their productivity fast enough to stay ahead, both because they cannot afford capital improvements, and because they need a significant proportion of their limited output to feed their families. Research and extension have been concentrated in the irrigation districts, and neither new crop varieties nor new systems appropriate for small holdings under highly variable natural conditions have been developed. Since 1966, marginal lands in rain-fed areas have been pulled out of production at a rate of .4 percent per year (Fernandez y Fernandez, 1978, p. 7).

Maize, which provides over 40 percent of both the calories and the protein in the average Mexican diet, is planted on between 50 and 60 percent of the total national cropland, over 90 percent of it under rain-fed conditions. It has been the basic staple of Meso-America since it was first domesticated over 7,000 years ago. Most of it is grown on farms of a few hectares, and is raised using traditional technology. Hundreds of varieties have been selected over the centuries which are adapted to very specific local conditions. The increased availability of hybrid seed, fertilizers, and other inputs increased national average yields from 700 to 1300 kilograms per hectare (from 11 to 20 bushels per acre) between 1950 and 1968 (Wellhausen, 1976, p. 129). Further progress is limited by several factors. The individual campesino's resources are limited; extension, credit and other services are inadequate; the parcels are too small to absorb modern inputs; and the tremendous variation in ecological conditions prevents the wide-spread adoption of improved varieties or more productive cultural methods.

Other basic crops grown under these conditions face these same limitations.

Until 1975, when the system was reorganized, official credit programs were scattered among several institutions following very different lending criteria. Credit was often granted as a form of subsidy or political reward; in 1962 it was estimated that 35 percent of the clients of the Ejidal Bank were totally incapable of repaying their loans. Between 1937 and 1973, the institution accumulated 4.7 billion pesos in bad debts, more than three times its total portfolio in the latter year (Aguilar, 1976, p. 35). The banks responded to this problem by limiting credit to small groups which they organized to pool their risks. This process, added to many other factors, has led to the development of favored groups within each ejido and rural community who dominate the local economies.

These patterns have been reinforced by the introduction of roads, electrical power, and other infrastructure into rural areas. As previously isolated communities came into direct communication with the national market economy, the traditional, localized systems of food production and distribution break down. The best soils are increasingly dedicated to relatively high-value crops, rather than to maize and other basic foods for local consumption. A campesino who used to grow enough maize to feed his family, and a little more maize to sell or barter for other basic commodities, now grows more valuable crops in his very limited surplus land to meet his increasing need for cash (Barraclough, p. 1977, p. 21). This surplus production has come under the control of the local or regional elite, either directly or through the merchants who control private credit and marketing services; the mechanisms of this process are discussed in more detail in chapter 4 (Stavenhagen, 1976, pp. 20-21).

Faced with a very unfavorable price structure, the small campesino does not have enough land or other resources to take advantage of the changing patterns of regional relative advantage by introducing modern cultural methods and specializing entirely in more valuable crops. He continues to grow at least some of his own food, but must seek off-farm employment to supplement his income. Work is scarce: 60 percent of the rural labor force was underemployed according to the 1970 census (Mexico, 5). The average productivity of labor in agriculture has been declining since 1970 (Scherr, 1978, p. 8).

POPULATION GROWTH AND THE STRUCTURE OF DEMAND

Mexican agriculture has not been able to produce enough food, especially basic grains such as maize and beans, to meet domestic demand since the late 1960s. The population of the country, which was 20 million in 1940, more than tripled to about 65 million by the end of 1977. If this rate continues, there will be 140 million Mexicans by the end of the century. Until a decade ago, both GNP and agricultural production were able to stay ahead of population growth. The daily per capita availability of calories increased significantly between 1945 and 1968; from 2,000 to 2,800, and protein consumption went up from 55 to 80 grams (Hewitt, 1977, Table 10). These average figures do not reflect the extremely unequal distribution of income and purchasing power among the population or the very different kinds of foods consumed by rich and poor people. These factors must be kept in mind when looking at the nature of Mexico's production crisis; the problem is much more complex than a simple race between population and food production.

Almost all countries experience a surge in population at a stage in their development usually explained in terms of the "demographic transition." In an agrarian society, a very high birth rate is balanced by a very high death rate, especially among infants and young children. One of the first results of the process of modernization is the improvement of medical and public health services, which rapidly bring down the death rate by controlling epidemic and endemic diseases. The birth rate, which is dependent on cultural tradition and on the need for children in rural areas to work on the farm and contribute to family income, comes down much more slowly. The population rises until urbanization, education, prosperity, and institutions such as social security eliminate the rationale for large families for the majority of the people (Poleman, 1964, p. 2).

The developed countries of Europe and North America went through the initial phases of this transition beginning in the 19th century. It coincided with the tremendous increase in demand for unskilled labor brought about by the Industrial Revolution. The terrible conditions of the urban poor during this period are well known. At different rates according to social class, race, and region, population has stabilized as the level of living has risen. Mechanization and increased productivity have permitted the agricultural sector to meet the rising demand for food in the urban areas.

Mexico, along with most of the Third World, did not begin to experience a population surge until the 1940s, but when it came, it hit very hard. The annual birth rate of 44 per thousand is very high compared to 37 in Brazil, 34 in India, and 14.4 in the United States. The death rate has fallen to nine per thousand, which puts the annual growth rate at 3.5 percent. Half of the people are under 16 years old and have not yet begun to have children. This means that even if average family size drops to the replacement rate, 2.3 children, by the year 2000, the population will not stabilize for another hundred years (Gordon, 1978, p. 4). Birth control programs are difficult to implement in a country where 40 percent of the people live in rural areas, where large families are still a real asset, and in a Catholic culture that stresses fertility.

Rapid population growth has not corresponded to a period of rapidly expanding opportunities. On the contrary, industrial development has been highly concentrated and relatively capital-intensive. Even if the economy returns to a post-war growth of 6 or 7 percent, at most 30 percent of the people coming into the labor force each year are likely to find full-time, permanent jobs in this sector (Gordon, 1978, p. 7). It has been estimated that four of the six million people who were unemployed or underemployed in 1970 were in agriculture (Poleman, 1978, p. 16). Unless jobs can be created much more rapidly than ever before, poverty is almost certain to increase.

Mexico already has one of the most unequal income distributions in the world. The poor, who have the least opportunities both in agriculture and in the cities, are increasing most rapidly while their share of the national income declines. The changes between 1950 and 1969, when the population increased by over 60 percent, are outlined in table 2.2.

Table 2.2. Mexico: Distribution of Family Income, 1950-1969 (percent of total by income group)

Income Group	1950	1958	1963	1969
Lowest 20 percent	6.1	5.0	4.2	4.0
30 percent below the median	13.0	11.7	11.5	11.0
30 percent above the median	21.1	20.4	21.7	21.0
15 percent below top 5 percent	19.8	24.3	24.3	28.0
Top 5 percent	40.0	28.6	38.3	36.0
Top 30 percent	59.8	62.9	62.6	64.0
Gini coefficient	.50	.53	.55	.58

Source: Cynthia Hewitt de Alcantara, "Ensayo Sobre la Satisfaccion de Necesidades Basicas del Pueblo Mexicano entre 1940 y 1970" (Mexico: Centro de Estudios Sociologicso, El Colegio de Mexico, 1977). Data from ILO study made in 1974.

According to the 1970 census, 72 percent of the families in Mexico made less than 1,000 pesos per month, or about 1,000 dollars per year at the exchange rate in that year. This figure is approximately the average per capita share of GNP. There has never been a reliable and comprehensive food consumption survey. Drawn from data based on very limited samples, figure 2.5 illustrates the general relationship between income group and the consumption of certain basic foods. It suggests that the great majority of the Mexicans ate less well in 1968 than in 1963. Inflation and rising population have almost certainly meant further deterioration over the past decade. The extent of actual malnutrition depends on how this concept is defined and measured. Estimates of the proportion of the population that cannot afford an adequate diet range from 28 to 40 percent (Gordon, 1978, p. 16; Kassorla, 1977, p. 41). Clinical symptoms of inadequate food intake among school children ranged from 10 percent in urban areas to 32 percent in the countryside in the early 1960s (Jiguera, 1977, table 4).

Poor people subsist on the same basic diet Mexicans have eaten for thousands of years: maize in the form of flat cakes called tortillas, beans, and chili - supplemented with an occasional piece of meat and other items. As one moves up the income scale, preferred foods such as wheat bread, meat, milk and other elements of the middle-class diet in the developed countries are substituted for the basic staples. Figure 2.5 indicates that in 1968 the richest group in the distribution consumed approximately ten times as much meat, 12 times as much milk, seven times as many fruits and vegetables, and three times as much wheat as the poorest; but only half as many beans and a third as much maize (Hewitt, 1977, pp. 31-32). In response to the greater buying power of the small, high-income portion of the population, the modern agricultural sector is placing more emphasis on high-value foods and animal feed crops than on maize and beans. At the same time, the government has kept the prices of these latter items down to control inflationary pressures, discouraging the use of marginal lands in rain-fed areas.

Mexico is being forced to import many kinds of agricultural products, but the greatest pressure is on these basic grains. Any significant redistribution of income in favor of the poorest half of the population, as was carried out in Chile during the presidency of Allende, would mean that poor people would use the additional money to buy more food. Unless the economic structure as a whole were entirely reorganized to meet the basic needs of the people, this would put impossible pressure on the systems of production and distribution, resulting in uncontrollable inflation and greatly increased food imports (Barraclough, 1977, pp. 22-23).

NEW GOVERNMENT PROGRAMS

The government, following the logic of capitalist development and strong pressures of political expediency, for many years put most of its resources into the modern irrigated areas where the natural and social

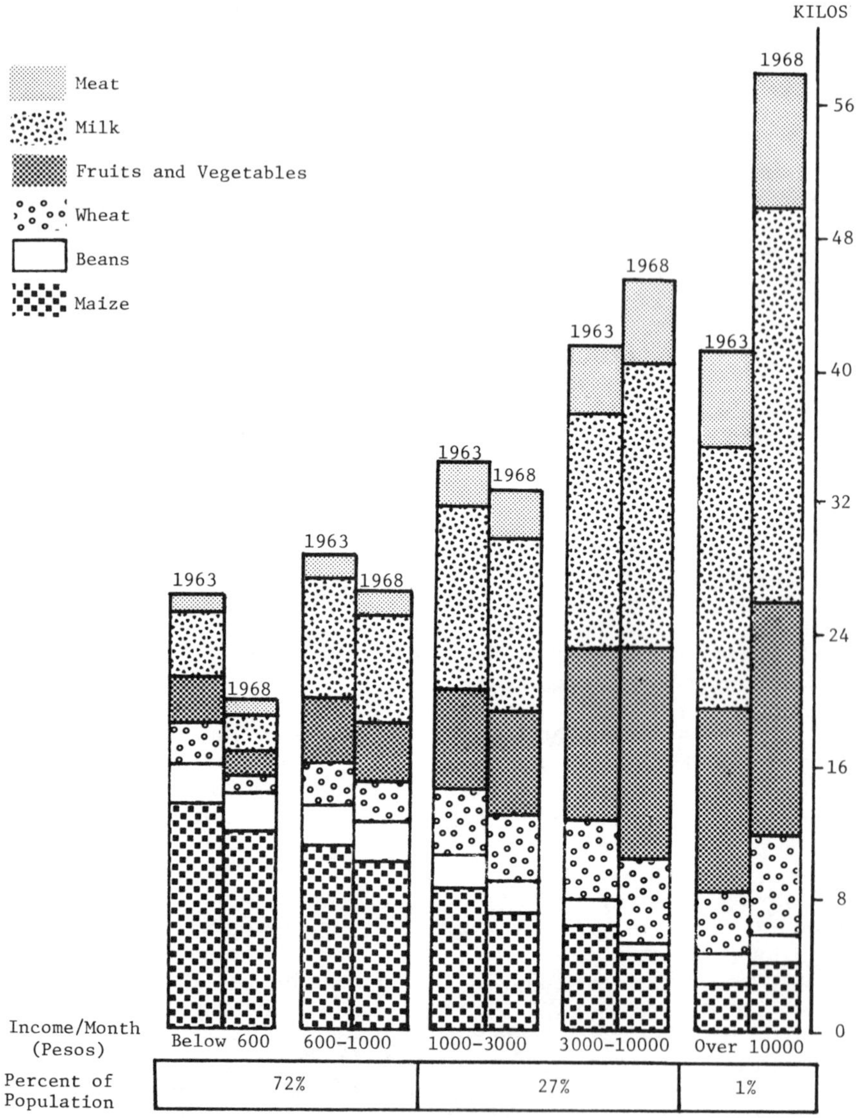

Fig. 2.5. Mexico: monthly per capita consumption of selected foods in kilograms, by family income group, 1963 and 1968.

Source: Cynthia Hewitt de Alcantara, "Ensayo Sobre la Satisfaccion de Necesidades Basicas del Pueblo Mexicano entre 1940 y 1970" (Mexico: Centro de Estudios Sociologicos, El Colegio de Mexico, 1977), pp. 40-41. Data from Banco de Mexico surveys.

conditions were the most malleable, where the return on invested capital was the greatest and the fastest, and where most of the direct benefits accrued to a favored few. By the early 1970s, it was faced with diminishing returns, increasing food imports, and a research, extension, and support structure which was totally inadequate to meet the needs of over 90 percent of the farmers.

A great variety of new programs were set up during the Echeverria administration (1970-1976) to grapple with the production crisis and to distribute services and income more fairly between the regions of the country and the sectors of the rural population. Special emphasis was given to the traditional areas, particularly to the ejidos. The agricultural portion of the budget was increased by over 700 percent, the credit system was reorganized, and new laws were passed to provide uniform standards. Scores of new agencies were created to deal with specific problems. The extension service was enlarged. The collective ejido was reintroduced as a way to reorganize fractionalized plots into large enough units to allow mechanization. Farm prices of basic foods were raised, and the costs of fertilizers and other inputs were subsidized. Financing was provided for rural storage facilities, small-scale industries, livestock improvement, and a wide variety of other programs. During the same period, the Mexican economy as a whole went into a serious decline. The short-term effects of these programs are difficult to measure, but the broad crisis in the agricultural sector continued to worsen.

The structural problems of the densely populated central highlands are deeply rooted, and cannot be modified easily. The attempt to collectivize the ejidos in these areas was virtually a total failure, primarily because it ran into the opposition of the local elites. Improved management could increase production in the irrigation districts, but it is difficult to create incentives for small, marginal improvements among people who are accustomed to very large profits and who may have many alternative investment opportunities. Given the existing political and social structure, Mexico is running out of easily-exploitable resources. As it moves to develop the more complex natural environment in the Southeast, it is attempting to find a new approach to satisfy the basic needs of its people.

3 The Tropical Southeast

The Southeast, as it usually defined, includes the states of Veracruz, Tabasco, part of Campeche, and the Gulf slopes of Oaxaca and Chiapas. The region is sharply divided between the tropical coastal plain and the highlands, where the highest peaks reach 5,000 meters. The rainy tropics lie below 150 meters (500 feet), in a band of varying width between the mountains and the sea, and include an estimated eight to ten million hectares of land with underutilized agricultural potential (Tomayo, 1976b). Nevertheless, the Southeast as a whole is by no means an undeveloped wilderness. Approximately 15 percent of the Mexican population lives there, over half of them on the coastal plain (nine million people, or 19 percent of the national population, were counted in the Plan Nacional Hidraulico's regions, VII, VIII, and IX in 1970; this area is somewhat larger than the Southeast as defined in this study). Two of the most important pre-Columbian cultures - the Olmecs and the Mayas - flourished along the coast as early as 800 B.C. Cortes first stepped ashore on the American mainland at Veracruz, and personally established the first sugar cane plantation. The region now produces about half of the sugar and a third of the cattle in Mexico. It has been the center of the national petroleum industry since the early part of this century. Recent discoveries of vast reserves, between 50 and 300 billion barrels, place the area in the forefront of future development plans. (Estimates of Mexico's potential oil wealth have been escalating rapidly. In early 1978, the Oil and Gas Journal put the country's proved reserves at 14 billion barrels. This term refers to the amount of crude oil which can be pumped economically at current prices from fields which have been accurately explored. This compares to 18.2 billion in Venezuela, 29.5 billion in the United States, and 150 billion in Saudi Arabia. In November of the same year, the C.I.A. put the proved reserves at 50 billion. Since 1973 PEMEX, the national oil company, has been exploring intensively and has discovered several important new fields. In his State of the Union address in September 1978, President

Lopez Portillo announced that the potential reserves - a term that includes some guesswork - were 200 billion barrels. A few months later, the Director of PEMEX pushed this figure up to 300 billion, which means that Mexico may be the world's richest source of petroleum [New York Times, November 29, 1978].)

Both population density and economic growth have been limited for centuries by the nature of the tropical environment. Epidemic diseases imported from Europe and Africa in the 16th century decimated the native population, and discouraged settlement by Europeans (Cook, Simpson, 1948; Gerhard, 1972). The abundant, seasonally variable rainfall has produced periodic flooding, and has leached the nutrients out of the soil. Communications were poor and unreliable until very recently. Commercial agriculture has consisted of extensive and inefficient cane operations, cattle ranches, and a few specialized plantations. Most of the rural people have lived in isolated communities, practicing traditional agriculture based on slash-and-burn techniques.

POSTWAR DEVELOPMENT PROGRAMS

The potential of the Southeast has been recognized for many years. In 1941, President Avila Camacho called for a "March to the Sea" to develop the coastal areas and relieve congestion in the central highlands. Disease control, roads, reclamation, and drainage would allow the organization of "a new type of tropical agriculture - which cannot be based on minifundia" (Poleman, 1978, p. 28). In 1947, the Papaloapan Commission was established to control one of the major river systems, build roads, and coordinate a broad range of services. The Grijalva Commission was created in 1949, but did not actually begin to function until the late 1950s. The planners involved in the early stages of these projects shared a broad vision of agricultural and social development.

In the 1950s and 1960s, a variety of public work programs, organized by these Federal Commissions and by PEMEX, had a significant impact on the region. Flood control, hydroelectric power, road construction, the expansion of urban centers, and sanitation received priority. The frontier was pushed back, many thousands of hectares of forest were cleared, and people moved in. Communities that had been isolated for centuries were brought into much closer communication with the national economy and culture. The elimination of malaria and other diseases brought down the death rate, and the population began to grow rapidly. Private investments in specialized crops such as pineapple and cocoa paid off well in certain areas (Scherr, 1978).

Agricultural output increased as new areas were brought under cultivation. Nevertheless, investments in efficient, highly productive systems were not made, because both public and private capital yielded a better return in other parts of the country. Very little serious soil or agronomic research was undertaken. Several promising agricultural and credit programs in the Papaloapan basin were discontinued for lack of funds (Poleman, 1964, Chapter 7). A relatively small pilot program in the Chontalpa region of Tabasco remained in the planning stage for nearly ten years, although it had been funded by international agencies (Alcazar, 1975, pp. 150-152). Sugar cane, the most important commercial crop, expanded only very slowly and the mills were allowed to deteriorate. Colonization programs were undertaken, but the new settlements were not provided with vital services amd did not prosper. Various perennial crops such as natural rubber were introduced, but basic technical and marketing problems were never solved. Visitors to the region in the late 1960s were struck by the deterioration of the physical constructions and the general listlessness of the programs (Warman, 1972, pp. 36-38).

NEW PRIORITIES IN THE 1970s

When the national agricultural crisis became acute in the early 1970s, the Southeast provided an opportunity to make an end run around the problems of the existing productive structure by taking advantage of unused potential capacity. By expanding the infrastructure that already existed, the government hoped to meet a variety of goals. The most important of these was to increase production of basic foods and other commodities, which were either being imported in increasing quantities, or were being produced at relatively high cost in the irrigation districts. It was also hoped that regional development would increase rural employment, narrow the disparity between the highly developed metropolitan centers and the backwards hinterlands, create markets for manufactured goods, and stem the tide of migration to the cities.

Intensive agriculture had not developed in the Southeast in four hundred years, in part because the population density was too low to support elaborate systems such as the paddy rice complex of Southeast Asia, and in part because national needs were met more cheaply in other areas. An attempt to alter the agroecological and social systems, which have developed over a long period of time in response to the natural conditions of a complex environment, is a very challenging task (Janzen, 1973).

SOUTHEASTERN AGRICULTURE

Slash-and-Burn Agriculture

Most of the cleared land in the Southeast is used for cattle ranches and sugar cane plantations. Most of the rural people, both in the traditional communities and along the forest frontier, practice some form of slash-

and-burn agriculture. The oldest and by far the most wide-spread farming system in the entire tropical world, it is also called shifting cultivation, swidden, and a variety of other names. Temporary clearings are made in the forest, crops are grown for a few years, and the plots are allowed to grow back in fallow. According to the broadest definition, 360 million hectares, or approximately one-third of the cultivated soils in the world, are managed in this way. Even in Southeast Asia, which supports the highest rural population density anywhere, 30 percent of the farmland is cleared on a temporary basis. This system provides food for 250 million people where good soils, fertilizers, and accessible markets are not available (Gourou, 1966, Chapter 4; Sanchez, 1976, Chapter 10).

A great variety of slash-and-burn systems from all parts of the world have been documented in detail (Conklin, 1963). The practice has been widely criticized for destroying the forest cover, as a cause of erosion, and as an inefficient use of land and labor. On the other hand, it is one of the few systems by which people can live on highly leached, infertile soils, and is based on detailed empirical knowledge. It has been a widespread practice in many parts of Mexico for thousands of years (Watters, 1968).

Each family selects a plot of forest, cuts down or girdles the trees, burns it over to clear it, and plants corn, beans, chili, and other seeds directly through the refuse and ashes with a long planting stick called an espeque. Burning volatilizes some of the nitrogen, carbon, and sulphur in the vegetation, but it has little direct effect on the soil organic matter. The ash provides considerable quantities of phosphorus, which is often in short supply in tropical soils.

Once the forest cover is removed, the soil is exposed directly to the sun and the rain. The soil temperature rises during the day, which increases the level of microbial activity and speeds the decomposition of the organic matter. The traditional farmers take care to leave a thick layer of mulch and establish a crop canopy within a few weeks, so the effect is not serious. Rapid planting with appropriate crops controls run-off and erosion, which is also limited by the abundant debris in the form of trunks, stumps, and ashes during the crucial period before the crop canopy forms.

There are very few weeds for the first year after clearing because the kinds of low-growing plants which like open sun cannot survive on the dark forest floor. By the second season, tough, fast-growing plants invade the plot and begin to out-compete both the crops and the farmers' efforts to remove them. Crop-specific pests and diseases are not a serious problem at first, because the density of any particular species in a natural tropical system is very low. As soon as a single field of one or a few crops is planted, their predators begin to concentrate as well. Since there is no winter season to break the cycle, pest populations build up very rapidly. The adverse effects of cultivation on soil structure are minimized by the practice of minimum tillage. When the labor involved in maintaining the plot is greater than that required to open a new one, the family moves on.

Various practices are used to maximize the utility of a particular clearing. In the first year, low-growing plants with a high nutrient requirement such as grains are planted. These are followed by taller, hardier crops such as cassava. Still taller species are then put in, both to mimic the natural succession of regrowth and to shade out the weeds. Multiple cropping systems of various kinds spread out the labor requirements through the year, reduce the need for weeding, reduce the danger of erosion, and slow the build-up of pests and diseases (Apple, 1972).

The length of the fallow period depends both on the fertility of the soil and on the population pressure on the land. When a field is abandoned, it is quickly over-run with grasses and vines. Within a few years, soft-wooded, fast-growing shrubs and trees take over and shade out the plants on the ground. This begins the secondary forest succession, which would require over one hundred years to reach the climax - a hardwood rain forest. This almost never happens, because after a period of years the grassy species are dead and the cultivators come back to clear the plot again. They often cut trees 50 to 100 centimeters above the ground (Hernandez X., 1959, pp. 24-25). If the roots are not damaged, the trunks will sprout, the secondary trees will reestablish themselves more quickly, and the fallow period can be reduced.

Most established communities are located on rivers. The recent alluvial deposits near the banks provide enough nutrients to be cultivated continuously. The fallow period gradually increases with the distance from the river.

As the Southeast has been opened up by roads and commerce, and as population has increased, the balance inherent in all well managed slash-and-burn systems has broken down. Fallow periods have been reduced and fertilizers have not been available, which has led to erosion and a decline in average yields over large areas. The stable equilibrium of a traditional system is very different from the practices employed by recent immigrants who follow new roads into virgin areas. Agriculture becomes a short intermediate stage between logging and cattle-raising. Slash-and-burn is a logical response to the tropical environment, but it cannot support more than a dispersed population or produce surplus food in any quantity.

Experimental attempts have been made to increase the regenerative efficiency of the fallow period and to substitute trees with economic value for the natural secondary growth. Strip farming and elaborate rotations have been developed (Greenland, 1975; Ruthenberg, 1976; Sanchez, 1976). These kinds of systems have not been widely accepted, partly because slash-and-burn farmers are almost by definition poorly educated and on the margins of society. At the same time, development agencies are under great pressure to increase production rapidly, and have little patience with elaborate and complex systems of agrosilviculture.

THE OLMECS AND MAYAS

This most basic system of tropical agriculture is usually associated with dispersed populations living in isolated areas, who produce little more than enough for their own subsistence. It is therefore one of the puzzles of Mexican prehistory that two of the earliest and most important indigenous cultures - the Olmecs and the Mayas - developed in the rainy tropics. Most early civilizations - in the Middle East, China, Peru, and the highlands of Mexico itself - centered around irrigated valleys in temperate regions. It is not easy to trace the transition from hunting and gathering bands to the formation of settled populations capable of art, science, and the other attributes of an advanced culture. Fossil remains of domesticated corn have been dated as early as 3500 B.C. (Rodriguez, Penalosa, 1975, p. 977). The Olmecs were living on the estuaries of several large rivers on the Gulf coast of what are now the sites of Veracruz and Tabasco by 800 B.C. They did not build stone temple complexes, and little is known of their culture and daily life. They produced sculpture of extraordinary sophistication, from monumental heads to delicate jade figures. The influence of their distinctive style, based on jaguar motif, is found all over Meso-America.

The sudden appearance of what was obviously an advanced people in a hot, rainy, and generally inhospitable climate has led some people to believe that the Olmecs were immigrants from Africa or some other part of the world (Rodriguez, 1977, p. 478). Most scholars, however, believe that an agriculture based on maize, beans, and a great variety of other crops developed simultaneously over many part of middle America, and that "the Olmec culture may have had a little bit more sun than the other cultural seedlings and so shot up more quickly" (Thompson, 1954, p. 51).

The achievement of the Mayas is even more extraordinary. Lowland sites on the coast of Guatemala date back to 1300 B.C. (Coe, Flannery, 1967). Between 300 and 900 A.D., a major civilization in the rain forest of southeastern Mexico and central Guatemala reached intellectual and artistic heights which no other in the new world, and few in the old, could match. The population in what is now virtually empty jungle reached several million. The remains of Tikal, Palenque, and scores of other religious centers display a high level of artistic skill. These people developed hieroglyphic writing, a literature, a numerical system which included the concept of zero, and an accurate calendar based on careful astronomical observations (Coe, 1966; Thompson, 1954).

Many attempts have been made to reconstruct the agricultural systems of the Mayas from the practices of their contemporary descendants, archeological remains, and the few ancient manuscripts that have survived. All available evidence indicates a very detailed knowledge of the soils, climatic patterns, and vegetation. Through an effective central organization, they were able to develop in harmony with the complexity of the tropics and produce a larger surplus than what this and similar areas are capable of today. Only priests and other specialists lived in the urban centers; the majority of the people lived on the land. Irrigation, terracing, and drainage were used in some areas, but slash-and-burn was the predominant system. Each family was able to produce a small surplus and donate several months of their labor each year to the construction and maintenance of the cities. Silviculture - the replacement of the naturally occurring trees with more valuable species - was part of the rotation system. The calendar may have been developed in response to the need to predict the first rains and other key moments of the year in a difficult and variable climate. The Mayas put an extraordinary emphasis on the concept of time; every day in the elaborate calendar cycle had a unique name and was worshipped almost as a god.

The Mayas have excited interest ever since their great temple complexes were rediscovered in the early nineteenth century. Many theories have been advanced to explain why the classic centers were abandoned to the jungle around 900 A.D. One of the most persistent hypotheses has been that the culture suffered an ecological collapse; population growth exceeded the carrying capacity of the naturally infertile soils and the available technology. This is very neat, but is based as much on speculation as on concrete evidence. Recent studies have demonstrated that both the history and the technology of these people was much more complicated; wars, political strife and other factors contributed to their decline.

One important lesson can be learned from the fact that the Mayas were able to build a major civilization in the jungle. Modern development programs tend to replace the complexity of the tropical forest with uniform stands of plantation crops, pastures, or trees. The fragmentary remains of this ancient people demonstrate that the intensive use of the multiple resources of the tropical ecological system can produce a significant surplus and can be sustained over long periods of time (Barrera et al., 1977; Drumond, 1968; Coe, 1966; Matheny, 1977).

By the year 1000, the center of prehispanic civilization had moved to the temperate central highlands around Oaxaca and Mexico City. When Cortes arrived in 1519, the tropical Southeast still supported a fairly large population in scattered centers along the coast and up the rivers. In addition to slash-and-burn, the various groups practiced permanent agriculture and produced a substantial surplus. Records of tribute paid to the Aztec empire include cotton, cocoa, rubber, and a great variety of other crops and natural products (Rodriguez, 1976; Rodriguez, 1977a, p. 479).

Diseases imported from Europe caused an extraordinary mortality all over the continent in the first generation after the conquest, but the effect was especially serious in the rainy tropics. By the 1570s, large areas were almost entirely depopulated (Gerhard, 1972, pp. 22-26). From a greatly reduced base, the traditional agricultural communities grew very slowly over the succeeding centuries.

THE HISTORY OF COMMERCIAL AGRICULTURE

Except for cotton and cocoa, most of the crops produced by the Indians declined in importance after the conquest. For the next 350 years, the Southeast was too sparsely populated, too isolated, and too unhealthy to produce more than a few commodities. In 1525 Cortes planted sugar cane and set up the first simple mill in Santiago Tuxtla, a volcanic area with good soils and a relatively healthy climate. African slaves were brought over to replace the natives and small mills were set up in several areas. Without the population base to support the semi-feudal colonial system of agriculture, livestock was one of the few alternatives open to the Spanish land owners. Cattle were quickly introduced, and by 1600 occupied most of the open land (Poleman, 1964, pp. 72-73). As long as the rivers provided the only transportation, settlement was limited to their margins. The vast expanses of unexploited land on every side provided no incentive to intensify production. The one large city was Veracruz, the only licensed port on the Gulf coast, through which all trade with Europe passed. Extensive and inefficient sugar and livestock enterprises have dominated the agricultural economy of the Southeast down to the present day.

After independence, in the late nineteenth and twentieth centuries, the construction of railroads and the expansion of export markets stimulated new investments in the region. The blockade on the Confederacy during the American Civil War encouraged the production of cotton. New factories were built and an industrial center grew rapidly around the city of Cordoba. The plantations prospered until the crop was displaced to the lower cost irrigated areas in the North, which were relatively free of the boll weevil. Coffee, tobacco, cocoa, and coconut plantations were expanded. New products such as mangoes were introduced. Sugar cane grew rapidly in importance and new mills were built, although yields were lower than in Morelos and other irrigated areas. Improved cattle and pastures were introduced (Rodriguez, 1977a, p. 492).

The unrestrained free enterprise liberalism that led to the rapid extension of the control of the haciendas over most of the agricultural land in Mexico during the presidency of Porfirio Diaz (1876-1911), faced special conditions in the tropics. The low population density limited the development of capitalist enterprises. Workers were brought in from other areas, some by the promise of high wages and others by force. Diseases and brutal working conditions were common. Crops such as sugar cane and tobacco with high labor requirements displaced the traditional farmers from the best land and threw the local food supply system out of balance. The hacienda stores became the only source of food, clothing, credit, and other necessities (Poleman, 1964, p. 61; Turner, 1969, p. 71).

Large areas remained empty. In the late 1800s, a series of laws were passed to encourage the colonization of unoccupied public lands, including large tracts in the tropics, by foreign and Mexican nationals.

Private companies were contracted to survey and distribute the land to settlers, in return for one-third of the area. The result was a tremendous wave of speculation by the members of the ruling elite. Between 1881 and 1889, companies under the control of only 29 persons acquired 27.5 million hectares, 13 percent of the total area of Mexico. Many Indian and other traditional communities lost their land through this maneuver (Silva, 1959, p. 116). Only very limited areas of the tropics were actually colonized. Many of the concessions were turned over to American companies, who took the mahogany, cedar, and other valuable wood out of the forests. The Standard Fruit Company set up large banana plantations in Oaxaca, Veracruz, and Tabasco, and Mexico was one of the world's largest producers in the 1920s. Serious outbreaks of Panama disease were too expensive to control and the Americans pulled out ahead of the application of the land reform laws in the late 1930s and early 1940s.

REVOLUTION AND LAND REFORM

The major events of the revolutionary period (1910-1920) were concentrated in the central and northern states. Nevertheless, the workers in the industrial cities of the tropics and on the railroads became important centers of trade union organization. Peasant leaders such as Vidal Tenorio led bands of campesinos to fight. The military situation was never very clear; towns and rail centers changed hands numerous times, and a great deal of blood was shed. The plantation owners, however, were able to continue and expand their operations.

In the succeeding decades, especially during Cardenas' administration, many large private holdings were distributed to the campesinos as ejidos. Most of the haciendas in the tropics had been formed within the previous 50 years. In areas that had long been occupied by homogenous Indian communities, such as the tropical portions of the state of Oaxaca, the lands were returned to their historical owners. Many ejidos, however, were formed of diverse groups of campesinos who had moved into the region only recently. By 1940, between 12 and 20 percent of the total area of the tropics had been divided into ejidos, about 30 percent of which was arable crop land (Whetten, 1948, p. 598). Most of the ejidos were located in the river valleys. Free from the obligation to work under very unfavorable conditions, the ejidatarios dedicated themselves to the production of basic foods on family plots using traditional slash-and-burn methods.

Many private cattlemen were able to exempt their holdings from expropriation by various legal maneuvers, and continue to manage extensive enterprises to this day. Most of them own businesses in the towns or in other parts of the country, where the return on invested capital is greater than from their tropical holdings. The Land Reform Law ties the maximum size of a cattle ranch to the number of animals it can support. This means that if the pastures were improved or other steps were taken to intensify production, part of the land could become eligible for expropriation.

Public lands in frontier areas were sold as private property until 1962. A new law was then passed that allowed the distribution of virgin territories to campesinos who had petitioned for ejidos in areas where land was not available (Ballesteros, 1970, pp. 171-9). These "New Ejidos" put the development of large tracts in isolated regions directly under the responsibility of the federal government. In many cases, few or no services were provided to the ejidatarios, who were not able to take possession of their grants and lost their rights after two years. The invasion of lands as they were opened up by new roads became an increasingly serious problem Widespread insecurity and conflicts prevented orderly development (Siemens, 1964).

Beginning in the early 1970s, the government began to organize ejidos into integrated, capital-intensive projects, both to relieve the pressure on the land and to increase production. The most important of these that preceeded Uxpanapa was the Plan Chontalpa in Tabasco (Alcazar, 1975; Barkin, 1978; Chao, n.d.). Uxpanapa is one of the last areas of virgin tropical rain forest remaining in Mexico. The Chinantec Indians who are being resettled are a traditional people who have been farming in the rainy tropics for over 1,000 years. If viable systems of intensive, mechanized agriculture can be developed, the model will be used to bring what is essentially Mexico's last frontier into the national economy.

4 The Chinantecs in Their Homeland

The Chinantec Indians live in the rainforest, on the steep slopes and in the narrow valleys which plunge down the northeastern side of the Sierra de Juarez, part of the mountain chain that forms one flank of the state of Oaxaca. The warm, moist rains that blow across the plain from the Gulf of Mexico during the summer are forced upward by this ridge, and the rainfall increases rapidly as one ascends. Within an air distance of forty or fifty kilometers, the average annual rainfall rises from 1,500 millimeters on the sea-level flat lands to over 4,000 millimeters on the upper slopes, where the ridges reach about 3,000 meters (Mexico, 22). The tremendous quantity of water drains back down through the forest in a series of torrential streams and rivers to feed the Papaloapan, Mexico's second largest river in terms of volume.

The varied and broken topography of Oaxaca has led to the formation and survival to the present day of over 15 linguistically and culturally distinct peoples, including the Zapotecs and the Mixtecs, the builders of Monte Alban and Mitla (Coe, 1962). The Chinantecs, separated from outsiders and among themselves by the rugged terrain, never built stone temple complexes or developed the arts and sciences of their more famous neighbors. They do, however, represent an example of culture that has adapted to the humid tropical environment, which twentieth century technology finds hard to balance.

A TRADITIONAL SOCIETY IN OAXACA

According to the most recent census, there are about 54,000 Chinantecs, over a quarter of whom speak only their own language, one of the Otomi-Zapotecan group of languages. The broad outlines of their history can be traced from the early Aztec and Spanish chronicles, through ethnographic notes taken by anthropologists since the 1930s, to the detailed censuses and studies made in the past five years to facilitate their indemnification and resettlement. One of the most fertile and populous areas in their domain will be flooded by the Cerro de Oro dam, and 37 percent of the Chinantec people will be moved, the majority of them to Uxpanapa. Although the climate is roughly similar in the two areas, much more is involved than a simple shift in location.

It has been 500 years since the Chinantecs were conquered by the Aztecs, and over 400 years since plagues introduced from Europe radically changed the Chinatecs' balance with their world. Especially in recent years, they have responded to many outside pressures as well as to the climate, soils, and vegetation of their natural environment. Nevertheless, many elements in the agricultural systems they use have changed very little. As farmers on a small scale, they know well every detail of their environment. Centuries of empirical experience have allowed them not only to support themselves, but to adapt new crops and new techniques with a high degree of efficiency.

The people are not hungry, but they are poor. They do produce a surplus for the urban market, but they receive only a small fraction of its value. Improvement of communications into the area, increased demand for consumer goods from the outside, population growth, and other forces have put the traditional system under stress. A very explicit goal of the Uxpanapa project is to radically change the society and the production systems of the Chinantecs; to take a group of "backward Indians" and turn them into productive citizens of modern Mexico. Although this particular program is a result of the construction of the dam and the need for resettlement, the government has undertaken a number of broadly similar projects in other areas of the tropical Southeast. The issues involved in the transformation of small scale agriculture, based on family labor, to large scale mechanized systems organized through highly capitalized cooperative enterprises, can in part be understood in terms of the history of the people.

THE ANCIENT CHINANTLA

Early Archeological Remains

Archeologists have concentrated their efforts on more spectacular remains in other parts of the country, but over fifteen major sites have been identified in the Chinantla, as the Chinantec area is called (Delgado, 1966). The earliest finds are pieces of pottery related to the Olmec culture of Tres Zapotes, which flourished on the coastal plain between 500 and 200 B.C. Later sites have been tied in with the influence of the Zapotec and Mixtec cultures, and the pottery, copper, and gold artifacts are identified with the basic chronology laid out at Monte Alban. For hundreds of years, the people have lived in the kinds of houses built of sticks and thatched with palm they still use today. Ancient population centers cannot be reconstructed accurately. Large sites with leveled terraces, walls, and simple tombs have been tentatively correlated with descriptions by Aztec and early Spanish chroniclers. There are several contenders for the site of the ancient Chinantec capital of Chinantla in the valley of the Valle Nacional River. The traditional date of its founding is 1110 (Espinosa, 1961, p. 75). The remains of Tlacuache are still clearly recognizable in the town of Paso Limon near Ojitlan.

The Aztec Conquest

The history of the region begins with its conquest in 1455-56 by the Aztez emperor Montezuma Iluicamina as part of his campaign against the Mixtecs (Villa Rojas, 1955, p. 67). He set up a garrison in Tuchteptl, the modern city of Tuxtepec, which means "hill of the rabbit." Records of at least two armed revolts in the second half of the fifteenth century demonstrate that the extraction of a considerable surplus in the form of tribute was much resented, and explains why the Chinantecs were so willing to take up arms with Cortes when he arrived. The names "Chinantla" and "Chinantec" are derived from the Nahautl word "chinamitl," which means "an enclosed space" (Bevin, 1938, p 9). Some commentators have suggested a connection with the "chinampas" or floating gardens of Xochimilco, but this seems less likely (Tamayo, 1976b).

The Tuxtepec garrison, on the Southeastern frontier of the empire, exacted tribute from the Mazatecs, Cuicatecs, and Popolucas as well as from the Chinantecs. Records exist of the goods that were shipped from two Chinantec towns. They included gold necklaces, gold encrusted shields, quetzal feathers, cotton cloth, rubber balls, vanilla, maize, beans, chili peppers, and especially cocoa in great quantity, which was so valued that the beans were used as a medium of exchange (Bevin, 1938, Chapter X). The Aztecs drafted local men into their army; the Chinantla battalion was said to number 100,000.

The Spanish Conquest

While a guest of Montezuma in Tenochtitlan, Cortes sent a party of men to the Chinantla to look for gold, which was panned from the rivers above Tuxtepec. Three Spanish soldiers liked the area so much that they deserted to set up a peanut farm. The search for gold was inconclusive, but Cortes was impressed by the reports of the Chinantec soldiers, especially by their long spears, which were "better than anything we have in Castille" (Castillo, 1939, vol. 2, p. 37). He sent Hernando de Barrientos to get some for his army. Bernal de Castillo described the scene as 2,000 Chinantecs entered the Spanish camp (Castillo, 1939, vol. 2, p. 60):

> (The Chinantecs) entered the city in great order, two by two. They carried long, thick spears with points that cut like razors....Each

> Indian carried a shield, and their banners were embroidered with feathers, and drummers and trumpeters played. An archer walked between each pair of lancers, and they were all whistling and shouting, saying "Long Live the King! Long live the King our Lord and Hernando Cortes in his royal name!"

Cortes was very pleased by this performance, generously gave them beads, and sent them home where they remained loyal to him in the subsequent wars.

Three days after the "noche triste," when Cortes' army was driven out of Tenochtitlan, the Aztecs in Tuxtepec fell on the Spanish garrison and killed them all: 72 Spanish men, six Castillian ladies, and over a thousand Tlaxcalan troops. In 1521 Gonzalo de Sandoval took the town back and burned the Aztec general alive. Spanish control was quickly extended over the whole region (Castillo, 1939). Cortes claimed the Chinantla for himself, but it was not included in the Marquesado del Valle, his enormous fiefdom composed of scattered holdings in Oaxaca, the Isthmus of Tehuantepec, Santiago Tuxtla, and elsewhere. Various encomiendas, or rights to tribute from the Indians, were granted to the Conquistadores (Gerhard, 1972, p. 302).

In the 1570s, a general census was ordered, and reports were sent back to Spain from all over the colony. Two of these from the Chinantec area have survived, the Relacion de Chinantla and the Relacion de Usila, both dated 1579. A Relacion de Ojitlan was written, but it has not been uncovered. (The two Relaciones are translated into English by Bevin, 1948, Chapter X).

In the form of answers to a standard questionnaire, these accounts provide brief information on the natural environment, the pre-Colombian and contemporary social organization, the tribute exacted, the crops grown, and some notes on the culture. Before 1519, the Chinantecs were divided into a number of villages, each of which was the center for a number of smaller villages. Each settlement had its own leader, or cacique, who was subservient to the chief in the town of Chinantla. These chiefs received tribute both in the form of various commodities and as a certain number of days of work each year. The Aztecs took their share of the tribute to Tuxtepec, where there was a regional court. Traveling officials applied the laws and settled local disputes. Aztec culture was extended through ceremonial occasions several times a year, which included fasting and human sacrifice.

The relaciones list a great variety of agricultural products, including most of the foods common in the area to this day: maize, beans, chili, zapote, sweet potatoes, cassava, pineapples, grapes, fish, deer, rabbits, armadillos, and a large number of wild plants and medicinal herbs. Cotton and cocoa were grown on extensive plantations along the banks of the river below Usila, which presumably extended down toward Ojitlan. A large number of dye plants, including indigo, were collected.

The people lived in villages on promontories, to save the best land for agriculture. Houses were built of sticks and thatch, as today. The temples were in the form of high towers "reached by one hundred steps," and the idols of a very large pantheon were kept in caves.

The natives of all America suffered an extraordinary mortality from diseases imported from Europe in the first generation after the conquest. Estimating the population of Mexico at the moment of contact is a lively source of scholarly dispute; figures range from three to twenty-five million (Cook, 1948; Gerhard, 1972). There is no question that epidemics of smallpox, typhus, measles, yellow fever and malaria, coupled with forced labor in the mines, killed the overwhelming majority of the people. The figures given in the relaciones are catastrophic; of the 16,000 families reported in Usila in 1520, 400 survived in 1579. In all of the Chinantla, where the army alone had numbered 100,000, less than 3,000 tributaries were counted (Gerhard, 1972, p. 303). The numbers may be open to doubt, but the basis of the native economy was utterly destroyed. The population necessary to keep large areas cleared and under cultivation was gone, the jungle rapidly encroached, and the survivors grouped together in scattered villages and eked out a subsistence with the kind of individual slash-and-burn agriculture typical of more primitive tropical cultures. The colonial economy depended on an organized native population capable of generating a surplus, which the Chinantecs were clearly no longer able to provide.

Some attempts were made to congregate the survivors into central towns; the people of the Santo Domingo valley were ordered into Ojitlan in 1571 by the viceroy. Nevertheless, encomiendas were not profitable and there were no mines. The Spanish generally shied away from tropical areas, and the narrow valleys and steep slopes were not suitable for the economic crops and the agricultural technology of the time. The Chinantla slipped into a backwater where it remained for 300 years. In 1743, only 97 subjects were listed in the town of Chinantla. The area was administered from the distant highland town of Teutila, in Cuicatec Indian zone (Gerhard, 1972, p. 300).

The missionary functions of the church were carried out by two groups, the Dominicans and the "secular" priests, who did not belong to any order. The Dominicans established centers in Teutila and Villa Alta, jurisdictions which include parts of the Chinantla. By 1570 there were secular priests living in Chinantla and Usila. The missionaries had great difficulty with the language; one became so frustrated that he ordered his flock to speak only Nahuatl, the Aztec language, which did not work out very well. Although the people were converted to Christianity, social services such as education were minimal. Saints and images, particularly the black Christ of Otatitlan, are much venerated to this day, and elements of the ancient religions have survived (Bevin, 1938; Gerhard, 1972).

The Traditional Society

The cultural hegemony of the Chinantecs has slowly broken down over the centuries. The people spread out into small villages, each of which depended on the productivity of the soil in the immediately surrounding territory. Slash-and-burn systems put very definite limitations on the number of people a given area can support. As families worked plots further and further from the central village, it became easier to set up small settlements near the fields than to walk all the way home every night. In some cases these peripheral settlements grew larger and more important than the original village. Many of the records that survive from the colonial period are legal documents of various kinds, and many of these are adjudications of the territorial disputes between groups that arose from this kind of movement. A number of stylized maps, or lienzos, have survived upon which the broad outlines of the history of this period have been traced (Cline, 1957; Espinosa, 1961).

Ethnologists have divided the Chinantla into four geographically distinct zones, each with its own dialect and customs. Since ancient times, the largest population groups have been concentrated in the two lowland valleys where the narrow bands of good alluvial soils allow intensive cultivation; the Valley Nacional and the Santo Domingo-Usila. These two groups, called the Chinantla Grande and the Chinantla Pichinche respectively, are traditional rivals. Over the centuries, a third group has moved out into the forests to the southeast of the Cajones River. They have hacked out migratory settlements near the margins of several smaller rivers. The fourth group, which has the smallest population of all, lives in the dry highlands in the rain shadow of the Sierra de Juarez centered around the town of Yolox. Chinantecs by linguistic affiliation, their culture is more closely associated with that of their neighbors, the Zapotecs (Weitlaner, 1936, pp. 523-552).

Social customs vary considerably between these regions, and many of them survive only as remnants of their former significance. The culture has tended to reinforce the independence of each small local community. The town of Ojitlan is divided into five distinct neighborhoods, or barrios. Until the 1930s, when the land was divided into ejidos, these lines extended into the boundaries of the territory, dividing land and rural settlements into sections like a pie. Each barrio was endogomous, meaning that people were not supposed to marry outside the group, and had its own ruling elders. A council of these men would meet periodically to discuss matters that affected the community as a whole. The elders played a religious role as curanderos, or healers. Other offices merged with those of the church, which acted as a focus for the community. When cooperation was necessary between individuals of different groups, an acuerdo, or special relationship, would be established. For special work, such as land clearing or house construction, a convite would be organized. Unmarried men served as topiles, which combined the roles of attending to the needs of the leaders and acting as a kind of police force, collecting taxes and maintaining order. Every Saturday, all of the men in the community were called upon to participate in a tequio, or fatiga, a community work project to maintain the paths and bridges and do other necessary tasks (Rankel, 1955; Weitlaner, 1951).

One of the most interesting handicrafts was the construction of suspension bridges out of vines, which were used to cross even very wide rivers. Tools, beautiful hand-woven embroidered clothing, dug-out canoes, nets, hammocks, and other basic necessities were made in each community. Unlike many of the Oaxacan peoples, the Chinantecs never developed a system of markets. Itinerant traders, most of whom were Zapotec Indians, until quite recently followed regular routes to trade goods from outside for produce. A high degree of cohesiveness and cooperation were maintained at the local level, but the Chinantla has not been a political, social, or economic unit for a long time (Rankel, 1955; Neitlaner, 1969; 1951; 1954; 1973).

MODERN HISTORY

Spanish laws protecting the rights of Indian communities were effective in isolated areas. The Chinantecs were not affected by the mainstream of Mexican history until the late nineteenth century, when President Porfirio Diaz encouraged foreign and Mexico interests to set up haciendas and develop the tropics. A native of Oaxaca, he launched the revolt against Lerdo de Tejada that began his career as a dictator with the Plan of Tuxtepec, which was proclaimed in 1876.

A railroad linking Mexico City and Veracruz was built in the 1870s. Spanish and German tobacco planters, many of whom were refugees from the Cuban revolution, set up plantations on the fertile flatlands in Valle Nacional and to a lesser extent in Ojitlan. Workers were imported from other parts of the country in large numbers. Valle Nacional became infamous as a penal colony, where prisoners were worked to death by barbarous living conditions and disease. John Turner, an American who visited the area in 1910 and wrote a famous account of the horrors he encounted, said that the Indians lived in the hills above the plantations, but does not discuss the effect on their own economy (Turner, 1969, Chapter 4).

Although the town of Ojitlan changed hands several times during the revolutionary period, the area was relatively peaceful compared to the violence elsewhere. An American banana company, Standard Fruit, set up operations in both valleys in the early 1920s, although the bulk of their plantations were further out on the plain. Several of the older Chinantecs interviewed in the course of this study worked as peons for the Americans. The banana economy was destroyed in the 1930s by uncontrollable outbreaks of Panama disease, and the foreigners pulled out as the Agrarian Reform distributed the lands. The company suspended its Mexican operations entirely in 1941. The tobacco planters also moved out in the 1930s and 1940s, and that activity has declined in importance, although high quality leaf for cigars is still grown on a small scale. Many of the Spaniards, Cubans, Germans, and mestizos who worked on the plantations and haciendas stayed on in the area to become merchants and cattle ranchers.

During the administration of Lazaro Cardenas, 90 percent of the land in the municipio of Ojitlan was divided into over 50 ejidos on the basis of the population living there at the time. Each settlement received a grant that gave each household an average of between five and fifteen hectares, depending on location and quality. Although most of the ejiodos have never been legally divided into parcels, the land has been broken into plots which have tended to stay in the same families in the two generations since their foundation. Although wealthy and powerful ejidatarios have been able to acquire the use of several parcels, illegal rental or sale to outsiders seems to have been minimal. A few ejidos were enlarged when there was still land available, but in general a rapidly rising population has been tied to a fixed and rigidly structured land base. The private property in the area consists of a few cattle ranches and a number of very small farms, or minifundia. The larger properties have acquired an economic importance out of proportion to their size and number.

Improved services have become available in the district in the 30 years since the Papaloapan Commission was established. An all-weather road from Ojitlan to Tuxtepec, the district marketing center, previously accessible only by river or on foot, was opened in 1954. Schools, health clinics, electricity, and other services have been introduced, although they have been concentrated in the town of Ojitlan. Over this period, the relatively isolated Chinantecs have been brought into a much closer inter-relationship with the dominant culture and economy.

In the 1930s, 71 percent of the Chinantecs could not speak Spanish and most of them were illiterate (Bevin, 1938). The population was not large in relation to the cultivated area and the variety of crops grown. The traditional values of the culture stressed community self-sufficiency. Needs from the outside were limited to the relatively few goods that could not be produced locally and were not prohibitively expensive given the high cost of transportation.

In the past forty years, the population of the three municipios that will be affected by the construction of the Cerro de Oro dam has grown as follows (Spindola, 1977, Chart N-2):

Municipio	1940	1950	1960	1970	Average Annual Growth
San Lucas Ojitlan	9,921	11,944	14,619	19,576	3.2%
Jalapa de Diaz	5,422	5,939	7,774	9,768	2.7%
San Felipe Usila	4,464	4,987	5,794	7,570	2.3%

According to the National Indian Institute (INI), the proportion of the Chinantecs who do not speak Spanish had dropped to 26 percent by 1970. Progress in education has been much slower: 30 percent of the population in the town of Ojitlan and 70 percent of the rural residents were still illiterate when the most recent census was taken (Lucero, 1974).

TRADITIONAL AGRICULTURE

The following sections on the agricultural, economic and social systems of the Chinantecs are based on interviews and on the following sources: Attolini, 1949; Barabas and Bartolome, n.d.; Bravo, Tamayo, and Jiminez, 1972; Carrion, Cruz, and Ortiz, 1977; Escanero, 1979; Guerere, 1970; Lucero and Avila, 1974, 1977; Martinez, 1970; Mexico, 19; Mexico, 27; and Rendon, 1973.

The Chinantecs of the Santo Domingo valley are faced with a very limited supply of good land, a highly variable and difficult climate, and a chronic shortage of capital. They grow a variety of crops - grains, beans, oil seeds, vegetables, and many kinds of trees - and gather products that are found wild in the forest. The basic productive unit is not the farm or the individual, but the family. The members of each household grow a combination of commercial and subsistence crops, work in other peoples' fields, make handicrafts for sale, and attempt to meet the needs of direct subsistence, necessary purchases, repayment of debts, and whatever small surplus they can accumulate. Intensive continuous cropping using fertilizer and other chemical inputs, short-fallow rotational systems, and extensive slash-and-burn are all used, often by the same family at the same time. The choice of crops and systems depends on the quality of the soils available and the kinds of resources the family owns, can borrow, or can rent. These choices are based on a detailed knowledge of the natural conditions and are limited by an economic and social structure that is heavily weighted against the great majority of them.

Geography, Climate, and Soils

The Santo Domingo is a major affluent of the Papaloapan. Including its own tributaries, of which the Rio Salado and the Rio Grande are the most important, it drains a basin of over 14,000 square kilometers. The Usila, or Santa Rosa River, joins it 30 kilometers above Tuxtepec. The lower valley of these two rivers, nearly 20,000 hectares of which will be flooded by the Cerro de Oro dam, support over a third of the entire Chinantec population. Most of the area is in the municipio of Ojitlan; smaller portions fall under the jurisdiction of Usila, Chiltepec, Jalapa de Diaz, and Valle Nacional. Young valleys on a geological scale, they have cut through Pleistocene and various sedimentary rocks. The land can be divided into four general zones, as shown in figure 4.1.

In some places the rivers run through gorges; where they do not narrow, strips of good alluvial soils have been laid down (I). The land rises, first to a relatively fertile plain (II), then to a peneplain made up of coarser materials (III), and finally to the steep hills above (IV). The width of the arable portion varies between two and ten kilometers, and the area as a whole lies between 20 and 70 meters above sea level.

The mean annual temperature is 25°C.; monthly means vary between

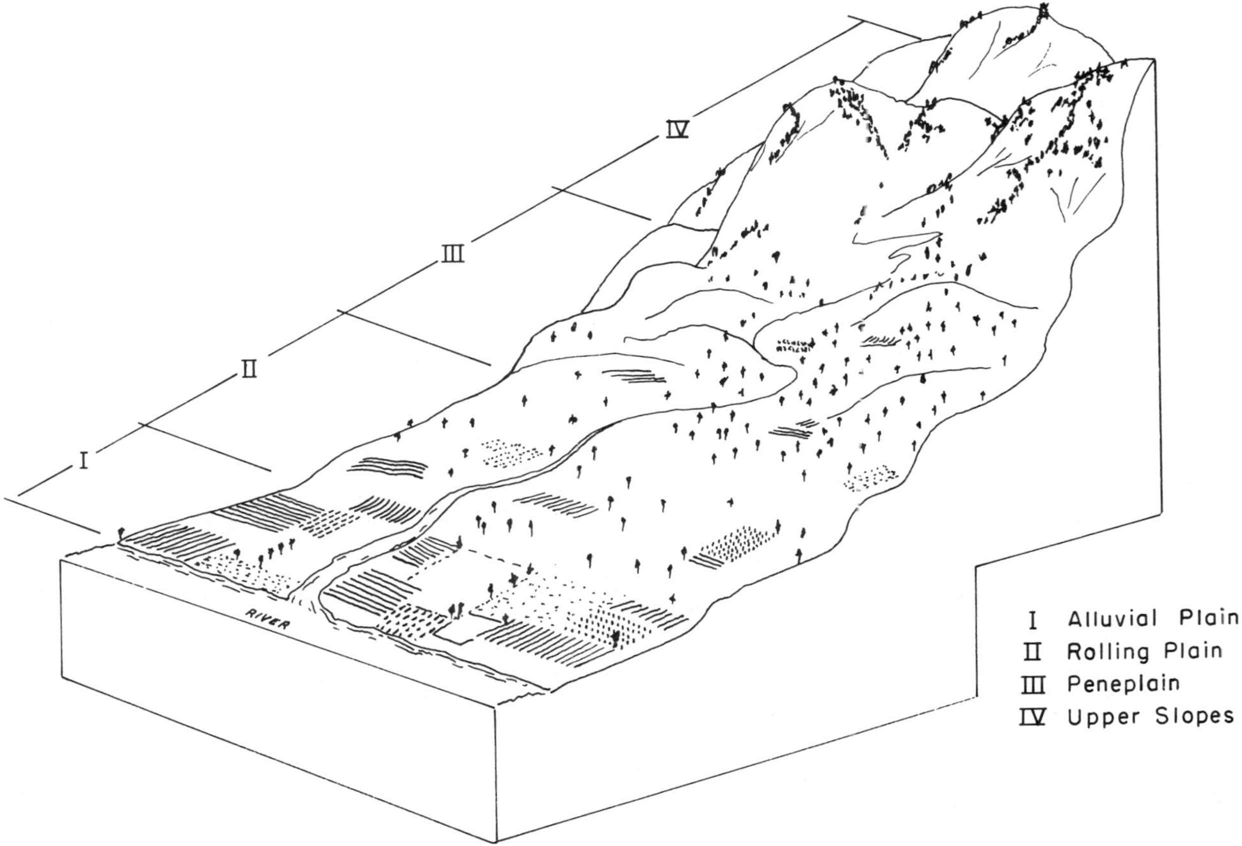

Fig. 4.1. Ojitlan: agro-ecological zones.

21° and 20°C. Average annual rainfall is 3,500 millimeters, 90 percent of which falls between the months of June and December. According to the Keoppen system, the climate is classified as Af, or humid tropical with a limited dry season (Mexico, 27). As in most areas of the tropics, both the total annual precipitation and its distribution through the year are highly variable; rainfall in any one year may vary as much as 1500 millimeters from the mean. Crop varieties with only modest yield potential under ideal conditions are chosen because they are capable of surviving these fluctuations.

The natural vegetation has been drastically altered over the centuries of human habitation. On steep slopes and in otherwise inaccessible areas, virgin rain forest of tall, evergreen hardwoods can still be found. The other areas not under cultivation in any given year are composed of a variety of secondary associations, depending on age, soil conditions, and other factors. The people have been forced by the necessity of supporting a relatively large population on highly variable soils to take careful note of qualitative changes over very short distances. An indigenous classification system divides the valleys into seven broad categories (fig. 4.2).

An individual campesino knows his parcel in much greater detail; he can easily point to as many as eight or ten significant qualitative differences in an area of two or three hectares. A comprehensive scientific soil survey has never been made of the valley. As the Chinantec classification suggests, the soils vary widely in their characteristics. A recent study in an ejido just outside of the area identified two sub-groups of latosols, gley soils, lithosols, and recent alluvium (Guerere, 170). The results of a land use study made in 1972 on the basis of aerial photographs are shown in table 4.1. It is obvious by any measure that very little arable land is available. An average family tills one to three hectares of good soil, plus another five or six in less favored areas. A Chinantec who has managed to acquire control over 20 or 30 hectares is a rich individual.

Cropping Systems

The six principal crops, whose relative importance are illustrated in figure 4.3, are grown in the various micro-ecological niches of the valleys with a complex combination of systems. Everybody grows maize and beans, the staple foods, on whatever land they have available. Commercial crops, especially rice and tobacco, which require significant investments in labor and purchased inputs, are grown on the best soils. The climate and rainfall distribution permit year-round cultivation. There are two principal cycles; the summer rainy season, or temporal, and the dry season, or tonalmil, which depends on the residual moisture in the soil supplemented by the relatively scarce rainfall. The first is limited to areas that are not too wet, the second to the much smaller area of soils with a high water-holding capacity.

The timing of agricultural activities to the weather is a major

I. *Huo Teh* (Moist Soil)

Alluvial soils along the rivers, suitable for two crops a year.

II. *Huo Cuah* (Hard Soil)

Well-drained soils on the plains, used for rainy season crops.

III. *Huo Seh* (Sandy Soil)

Well-drained soils on the peneplains and lower slopes, used for rainy season crops and root crops such as cassava.

IV. *Huo Neh* (Yellow Soil)

Poorly drained soils at the base of hills and around springs. Used for early maize crops and some rice.

V. *Huo Yin* (Red Soil)

Lateritic soils found on higher land. Used for pastures, summer corn, and heavily mulched backyard gardens.

VI. *Huo Mah* (Black Soil)

Thinner soils on steep slopes, mostly in forest. Used for tobacco seedling nurseries and for slash-and-burn cultivation.

VII. *Huo Hmeh* (Forest Soil)

On very steep slopes. Used for wood, barbasco, and other natural forest products.

Fig. 4.2. Ojitlan: Chinantec soils classification.

Source: Alvaro Lucero and Simon David Avia, "Proyecto Cerro de Oro," manuscript, Mexico, 1974.

Table 4.1. Future Reservoir of the Cerro de Oro Dam: Land Use Study (area in hectares).

Use	Area	Percent of Total
	(hectares)	
Both rainy and dry season crops under continuous cultivation	516	3
Rainy season crops under continuous cultivation	3,040	16
Crops in areas managed with the short fallow system	396	2
Areas in short-term fallow	2,063	10
Crops in areas managed with the long fallow system	2,086	11
Secondary growth less than two meters tall	3,435	18
Secondary growth 2-5 meters tall	1,803	9
Primary and secondary growth over 5 meters tall	3,866	20
Pastures	1,467	7
Pastures infested with woody growth	320	2
Other uses	22	1
Rivers	457	2
TOTAL	19,471	100
Total under cultivation	6,060	31
Total in pasture	1,467	8
Total not in production	11,944	61

Source: Mexico, Secretaria de Recursos Hidraulicos, Subsecretaria de Planeacion, Estudios y Proyectos, S.A., "Estudio Agrologico Especial del Vaso de Cerro de Oro y Las Areas de Ojitlan y San Felipe Usila, Oaxaca," mimeo, 1972.

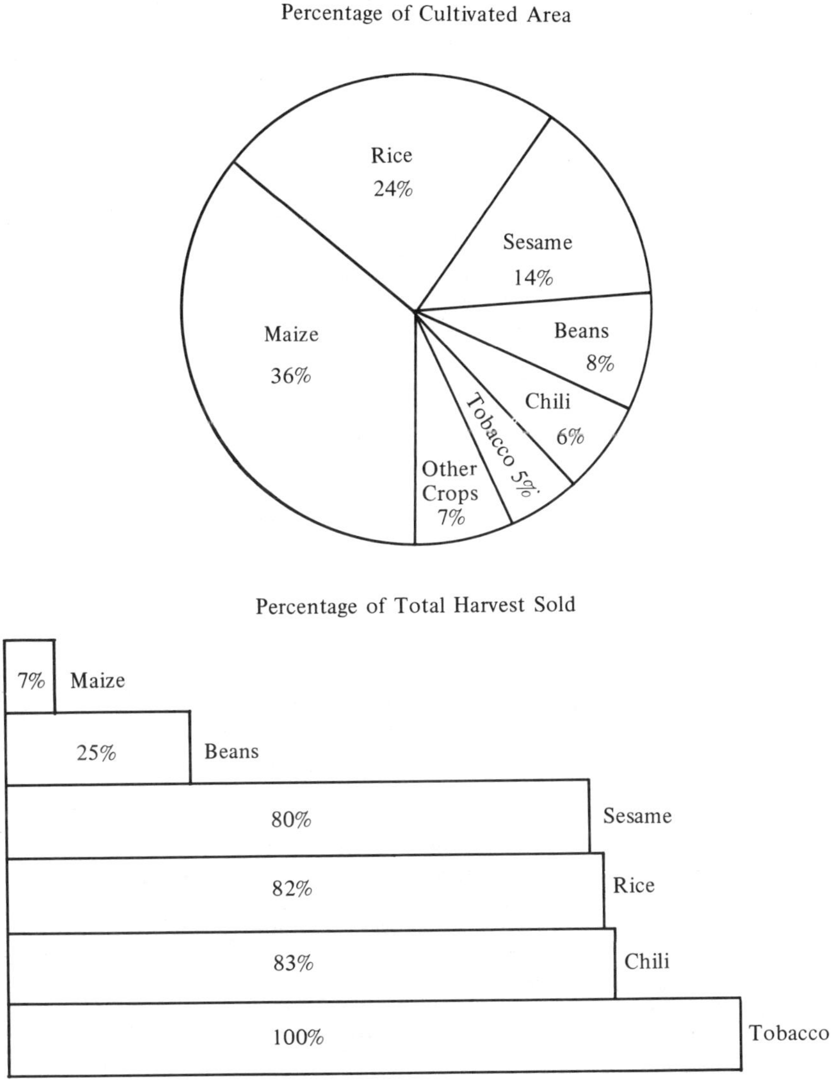

Fig. 4.3. Ojitlan: relative importance of principal crops.

Sources: Area distribution from A. Lucero and S.D. Avila, "Proyecto Cerro de Oro" (manuscript, Mexico, 1974). Marketing information from F. Escanero Gonzalez, "Reacomodo de Campesinos Afectados por la Construccion de la Presa Cerro de Oro en Uxpanapa, Veracruz" (Mexico: Tesis Professional, Facultad de Ingenieria, UNAM, 1977).

preoccupation of farmers everywhere. The pre-Colombian calendar of 18 months and 20 days plus an extra "month" of five days survived until recently in many parts of the Chinantla apparently because of its usefulness in coordinating the cropping cycles (Weitlaner, 1936; 1946). There are two especially crucial moments in the year; the beginning of the rainy season, on which depends the timing of the spring planting, and the beginning of the dry season, on which depends the time available for getting the summer crop off, preparing the soil, and replanting. No two years are ever alike, but figure 4.4 outlines the distribution of tasks through the year. It is important to note that there are very few weeks when there is nothing to do.

Almost all of the Chinantecs have different plots in different ecological zones, and they have developed different systems to take advantage of their special characteristics. The alluvial bands along the rivers are continuously cropped. Although it was reported as recently as the 1930s that traction animals were unknown in the Chinantla (Bevin, 1938, p. 43), steel moldboard, wooden, and small disk plows drawn by oxen are now commonly used in these areas. Tillage improves the structure of the heavy soils and speeds up the turn-around to permit year-round cultivation. The land is plowed twice, once in the spring and again in the fall when a week or more of sun has made the soil dry enough to work. The seedbed is leveled by dragging a log over it. In some cases the land is furrowed. These areas are generally planted with rice in May or June and with maize, tobacco, or chili in the fall and winter.

The higher lands corresponding to zones II, III, and IV in figure 4.1, are not tilled, but are cultivated using slash-and-burn techniques. The areas on the upper slopes are left in fallow for at least four years. A plot is marked out a few months before planting. Its size depends on the labor that will be available through the season; two and a half hectares is the most an average family can handle. Most of the trees are cut down, and a border is sometimes cleared to prevent the fire from escaping. The plot is burned over during the dry season. Summer maize is the most usual crop. On steep slopes, the man walks down the hill, pushing his espeque into the ground, dropping a few seeds into the previous hole, and covering them with his foot, all in one easy and continuous motion. It would be awkward to work sideways along the contour. Planting in vertical rows also facilitates drainage, but leads to erosion when an entire hillside is in production at the same time.

The rolling plains in the intermediate zone of the valley are left in short fallow, or barbecho, for one to four years. The length of the fallow period has been considerably reduced in recent years. The ejidatarios know that this prejudices yields and leads to erosion, but it is unavoidable, because of the pressures on the land. The succulent secondary vegetation, or acahual, is cleared with a machete and is either left on the field as a thick layer of mulch or is burned.

The technology varies little from ejido to ejido, and is typical of southeastern Mexico as a whole. Mechanization of any kind is limited to a handful of private farms. Only about a quarter of the members of

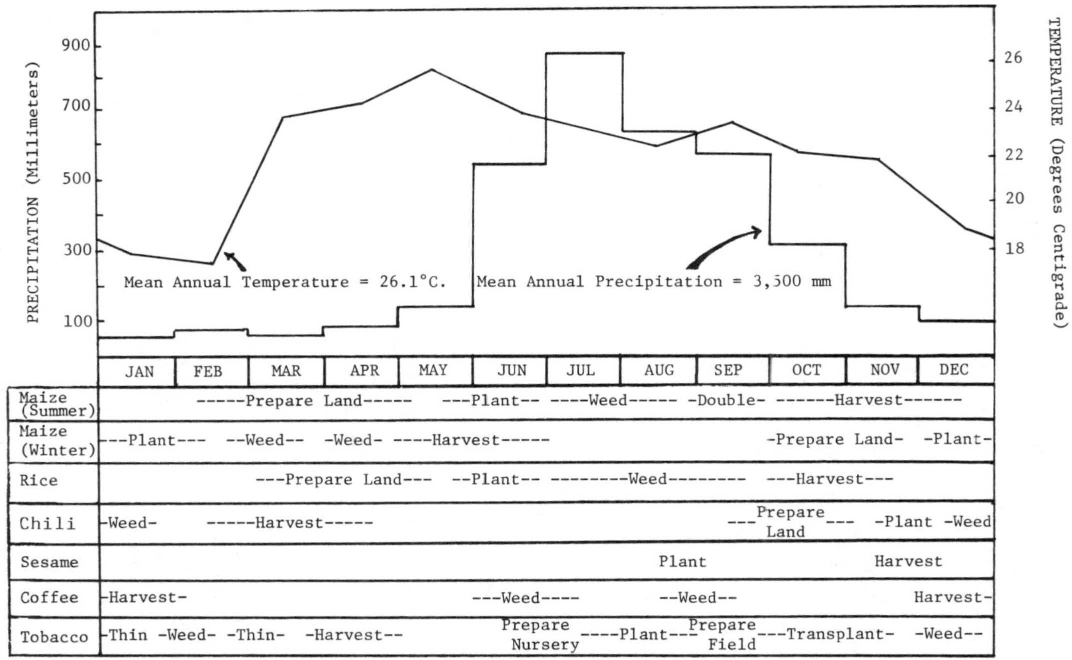

Fig. 4.4. Ojitlan, Oaxaca: generalized outline of agricultural year.

Source: Climatic data from Mexico, Secretaria de Recursos Hidraulicos, Comision del Papaloapan "Boletin Jidrometrico #20," 1974. Cropping cycles from interviews.

any ejido own teams of oxen and the equipment that goes with them; those who do, hire out members of the family to do custom work on other parcels. Chemical inputs have been introduced, but their use varies widely between ejidos and individual campesinos. They are used very selectively in all cases. According to the local people, a "good" soil produces high yields without fertilizer. In less favored areas, or where adequate fallowing is not possible, fertilizer is necessary, although most ejidatarios believe that it damages the land. In a small plot, it is possible to watch the stand carefully and fertilize visibly suffering individual plants.

Herbicides are used more widely, especially to reduce the fallow period in untilled areas. The invasion of grassy weeds, particularly Simsia amplexicaulis, often forces the abandonment of a plot before the absolute decline in soil fertility becomes apparent (Mexico, 27). In the last few years, inflation and the availability of relatively highly paid work in Uxpanapa has driven up the cost of labor. Some people have been using herbicides instead of paying peons to weed. Insecticides are used on tobacco, which is grown with government credit, and to some extent on rice.

The constant high humidity and inadequate facilities make storage pests a very serious problem. Many families are forced to sell maize at low prices before it spoils, and then turn around and buy enough to eat on credit against the next year's harvest.

As would be expected in an area of variable soil conditions and very small units of production, the details of the management systems are not uniform. Families vary in their size; in their access to land, credit, oxen, labor, and other inputs; and in their adaptability, energy, and skill.

Crops for Home Consumption

Maize is the principal food staple in the area, as in all of Meso-America. Two crops a year are grown under a wide range of conditions. Yields vary between two and three tons per hectare, and average between 1.5 and two tons. There are several native varieties for both the winter and the summer seasons. Hybrids are used by groups that receive credit from the official bank, but the ejidatarios prefer the native types, both because they can select and save their own seed, and because the yields are more reliable. Table 4.2 outlines the general cash and labor requirements for the crop. The actual costs vary considerably depending on the fertility of the soil and the amount of labor provided by the farmers' families.

Black beans and a few other varieties are grown on a small scale by most families. The seed is sometimes planted in the maize fields in August and September. This practice economizes land by allowing the two cycles to overlap, and saves time and labor. Some campesinos do not find this relaying convenient, and sow beans in monoculture later in the fall. The yields vary from almost nothing to about two tons per hectare.

Table 4.2. Ojitlan: Generalized Outline of the Cultivation of Maize on Untilled Land

Task	Operation or Input	Tools	Season	Man-days per Hectare	Unit Cost*	Cost per Hectare
Land Clearing (long fallow)	Cut trees and break up trash to facilitate burn	Ax, machete	Feb.-Apr.	20-30	30	600-900
Land Clearing (short fallow)	Cut secondary growth	Ax, machete	Mar.-Apr.	15	30	450
Land Clearing (no fallow)	Clear trash in plot just harvested	Machete	May	5	30	150
Burn	Make firebreak (not universal)		May	1-4	30	30-120
Seed	11 kilos per hectare (usually saved)				10/kg.	110
Plant	By hand, with stick	_Espeque_	May-June	5-8	30	150-400
Fertilizer	Variable small quantities				2.60/kg.	variable
Fertilize	Apply plant-by-plant by hand, as needed		Late June	2-3	30	60-90
First Weeding	Remove weeds and loosen earth	Machete, adze, hoe	30 days after planting	5-7	30	150-210
Herbicide	Variable quantities				70/liter	variable
Apply Herbicide	By hand, on weeds in row			1-2	30	30-60
Second and Third Weeding	The longer the plot cultivated, the more serious the weeds			6-12	30	30-60
Dobla	Bend over mature ears to dry; plant sesame		Aug.-Sept.	2-3	30	60-90
Harvest			Oct.-Dec.	5-8	30	150-240

*Price per man-day unless otherwise noted, in pesos in 1976.

Source: Interviews.

A variety of vegetables, fruits, and other useful plants are grown on a small scale, primarily in backyard gardens. Small quantities of cassava, sweet potatoes, mangos, bananas, oranges, coconuts, and other products are sold, especially by people who live relatively close to the main road. There is no organized system to reach distant markets, and local prices at harvest time drop to a point where it is not worthwhile to transport them very far. A study by a botanist in an ejido near Tuxtepec identified nearly ninety edible plants in the garden around a house, eighteen trees used as a source of wood, six species used to make furniture and tools, and fifty-five wild and domesticated plants for medicinal purposes (Alfaro, 1970). Game has become scarce in the past twenty years, but deer, armadillos, and other wild animals provide an occasional source of meat. Every household keeps pigs, chickens, turkeys, and other domestic species that eat residue and surplus. Many of the villages are built on high land on inferior soils, and the free-running animals provide the fertility of the gardens.

Commercial Crops

As the region as a whole developed in the 1950s and 1960s, the production of commercial crops became increasingly important. By the 1970s, nearly half of the cultivated area was dedicated to four crops grown primarily for sale, of which the most important is upland rice (figure 4.3). This grain is grown under a variety of systems in other parts of Oaxaca and Veracruz; in these valleys, its culture is confined exclusively to the tilled alluvial soils along the rivers. It is planted in May or June by espeque in clumps. Hardy, tall varieties such as Lira and Morada Criollo are the most common. When IR-8, the first of the "Green Revolution" dwarf varieties, was marketed locally under the name "Philippine Miracle," a few campesinos tried it, but it proved susceptible to disease and difficult to harvest.

The fields are cultivated once or twice with a single ox trained not to trample the plants. An additional hand weeding is often necessary. Fertilizer is used sparingly on individual clumps to produce an even stand. Insecticides and rat poison are applied if the pests are serious and if the resources are available. The grain is harvested by machete and threshed in the field by beating the panicles against a bamboo frame or against the edge of a metal tub. A large family may be able to handle most of the work up to the harvest itself; help is universally hired to gather the grain, a job that requires about 15 man-days per hectare and must be accomplished rapidly.

Yields vary between one and three tons per hectare. The harvest is usually sold in advance to secure the credit; the merchant arranges for mules or a truck to pick it up at the field. An outline of the stages of production is given in table 4.3.

The cultivation of tobacco requires a great deal of time, labor, and highly specialized skills. The crop was introduced into the area in the last century by foreign capitalists. Credit and marketing are now

Table 4.3. Ojitlan: Generalized Outline of the Cultivation of Upland Rice on Tilled Land

Task	Operation or Input	Tools	Season	Man-days per Hectare	Unit Cost*	Cost per Hectare
<u>Chapeo</u>	Clear and burn trash and residue	<u>Guataca</u> and machete	April-May	4-6	30	120-180
Plow		Oxen and plow	May	3-4	75-125	225-500
Harrow		Oxen	May	2-4	75-125	150-500
Level		Oxen and trunk	May	1-2	75-150	75-250
Plant	60-80 kgs. seed				5-6/kg.	300-480
	200-300 kgs. fertilizer				2-3/kg.	400-900
	Plant	<u>Espeque</u>	June	8-12	30	240-360
Weed	Herbicide (2 liters)		June-July		75/liter	150
	Cultivate	Single ox				
	Hand weed			10-20	30	300-600
Harvest and Thresh			October	15-20	30	450-600

*Price per man-day unless otherwise noted, in pesos in 1976.

Source: Interviews.

administered by TABAMEX, a federal agency. The leaves are harvested in the spring, cured in the sun, and dried in special sheds. They are graded into different classes, each with its own price. If the curing and selection are not up to standard, the campesinos receive the lowest possible price for the whole crop, which inevitably means a loss. The production of high quality leaf requires 27 separate hand operations, and a much higher investment in purchased inputs than any of the other crops grown in the area.

Sesame grows quickly and requires relatively little investment or care. To save time and labor, it is usually broadcast in the maize fields in August. It is not fertilized or weeded and is harvested along with the grain. Yields average between 300 and 800 kilograms per hectare.

Several varieties of chili peppers are produced as a cash crop. The labor requirements are very high - as much as 40 man-days per hectare for the harvest alone - so most families plant less than a hectare. Yields are as high as eight tons per hectare, but the market is subject to wide price fluctations. Chili can be very profitable, but it is risky.

Coffee is grown on the upper slopes. It is not the ideal elevation, and the Chinantecs have not been able to provide the careful husbandry needed to insure high yields. The price has been unfavorable for long periods. It is a marginal resource that provides a little extra income. In many ejidos, a few individuals control most of the plantations, the others own a few dozen bushes each. Although it is the only perennial crop systematically exploited on any scale, the plantations are overshaded, inadequately weeded, and only occasionally pruned. Yields vary from less than one to four kilos of dried coffee per bush.

Barbasco (*Dioscorea composita*) is a wild yam that grows in the forest. Steroids extracted from the tuber are used in the production of birth control pills. Before this use was discovered, the Chinantecs put the extract into streams to poison the fish. Until a few years ago, private brokers for the American laboratories paid low and unstable prices. In the early 1970s, PROQUIVEMEX, a government agency, took over buying and distribution. Sales to the Americans were embargoed in 1976 and a higher price was negotiated. Specific buying seasons have been established to limit supply and conserve the resource.

THE IMPACT OF MODERNIZATION

In the two decades since the road was built into Ojitlan in the early 1950s, the economic and social structure of the Chinantecs has been strongly affected by the national market and culture. A generation ago, the people grew cotton, spun thread, and wove beautiful embroidered clothing. The displacement of cotton by more profitable crops now requires the purchase of manufactured thread. The native *huipiles* are now so expensive to produce that few women can afford to wear them; they are sold back to the merchants who make a large profit on them in the tourist markets in Oaxaca and Mexico City. The people produced a beverage called popo out of cocoa and maize; its production for festivals was traditionally the province of a special group which thereby asserted its position and status in society. This continues on a minor scale, but the major organizing force behind the festivals is now the local distributors of the national beer companies. They provide tents, tables, chairs, musicians, and of course beer, which must be purchased, which requires cash, which is obtained as credit.

Many examples of the substitution or introduction of goods from the outside could be given. The only way in which the Chinantecs have been able to carry the credit loans imposed by these purchases has been to substitute commercial crops such as rice for subsistence crops such as maize. They have combined their ancient technology with new elements that are appropriate to their scale of production. Oxen, plows, fertilizers, and other inputs have been introduced.

According to census figures, the area under cultivation in the municipio of Ojitlan more than tripled between 1960 and 1970 (Rendon, 1973, Chart 6-3). Crop-reporting systems improved during this period, so part of the increase is simply due to a more comprehensive count. Nevertheless, both the area and especially the proportion of the harvest that is marketed have increased significantly since the road was built.

The best land is now dedicated to a limited number of crops for sale, which require hired labor and other purchased inputs to produce. Maize and beans have been displaced to the more delicate and inferior soils on the slopes. Increased intensity in these areas has required the reduction of the fallow periods; the campesinos must buy inputs to fight weeds and to maintain fertility. These forces, combined with the increase in population, are pushing the agricultural use of the area beyond its ecological limits, resulting in erosion and a general decline in fertility.

Gente de Razon and Paisanos

The economic system in the area, like that in many of the Indian communities of Oaxaca, is based on a highly unequal social structure defined by racial distinctions. It has been called a caste, rather than a simple class, system (Beltran, 1967). The ancient town of Ojitlan is located on a hill overlooking the Santo Domingo River and will survive on a peninsula when the Cerro de Oro dam is completed. Its population in 1970 was a little over 3,000, 17 percent of the municipio. The church, the municipal offices, and the school enclose the main square at the top of town. The single main street is lined with small stores.

Roads out to a few of the ejidos are passable by trucks except in the very wettest times of the year, but most of the settlements are accessible only on foot or by river. Only five of the over 50 villages in the municipio have a population of over 500. In recent years, schools, electricity, and other services have been provided in some of the ejidos by the Papaloapan Commission, but most of the modern improvements have been located in the town. Except for small stores with a limited stock of basic necessities, soft drinks, and beer, commerce is concentrated in Ojitlan.

The merchants are Spanish-speaking mestizos, many of whom are members of families who worked on the haciendas in the first part of the century. In a terminology which dates back to the early years of the colony, they are called _gente de razon_, or "people of reason." The Indians, over 20 percent of whom still speak only their own language, are called _paisanos_, an implicity pejorative term.

Informal Credit

It has been the local merchants - the gente de razon - who have benefited most from the increased production of commercial crops. A study by the Bank of Mexico indicated that the average per capita income of the campesinos in the municipio declined during the 1960s, and the income of the people in the service sector increased (cited in Rendon, 1973, Graph 6-3).

The merchants control the economy of the area in a number of ways, the most important of which is credit. They enter into life-long relationships with a client group of Chinantec families. They know everybody intimately, and often act as godparents for the children. This sets up a close personal tie known as _compadrazco_. They provide credit for goods of many kinds; seeds, fertilizers, food, clothing, liquor; and cash for emergency needs such as illness, unexpected trips, and weddings. They take on almost all of the short-term commercial risks. They provide the inputs and the cash to pay the labor, they send their mules or trucks out to the fields to pick up the harvest, and they take responsibility for transportation, storage, and marketing.

In return, the Chinantecs are obligated to turn over a large share of the harvest to cover their debts, and to provide political support to the merchant or his faction. To sell elsewhere would not only be an affront to an old friend, it would invite retaliation by force. Few people get hurt for stepping out of line; the rules of the system are well understood by everyone and are more subtly enforced. The ejidatarios do not hold title to their land and cannot put it up as collateral; if they could, they would have lost it long ago.

As long as the value of production in the area exceeds the costs of the inputs plus whatever is consumed, there is no way that the merchants as a group can fail to make money in the long run, whatever may happen in a particular case or year. The private lenders charge usurous rates and acquire life-long control over their clients, but their credit is available as needed in small quantities for a broad spectrum of requirements, including consumption.

Much of the surplus value of the increased agricultural production has been transferred, not only out of the hands of the Chinantecs, but out of Ojitlan. A detailed study of this process has never been made, and there are very few reliable statistics. The rice mills in Cordoba have used the region as a source of supply for years by financing loans through the merchants, who are also backed by wholesalers and other outside interests. Tuxtepec, the regional center, has benefited from the development of the surrounding area and grew very rapidly in the late 1960s and 1970s. Cattle ranches in the coastal plan have expanded and have become increasingly capitalized. The merchants have had many investment opportunities outside of the municipio. As in all of rural Mexico, the benefits of growth have been very unequally distributed.

The Social Structure of the Rural Communities

The capital that has remained in the area in the form of cattle, oxen, farm implements, improved housing, and so forth, has been increasingly concentrated within the Indian communities themselves. The Chinantecs who live in the town of Ojitlan and work in commerce or for the government have benefited the most. Houses built of concrete block, just like those of the gente de razon, televisions, and other consumer goods are the marks of their status. The people in the countryside can be roughly divided into the following five groups:

1. The mestizo ranchers.
2. The rich and powerful ejidatarios who have been able to acquire control over relatively large areas of land and cattle.
3. The ejidatarios with parcels who also own oxen and other tools they rent out.
4. The ejidatarios with parcels who depend on their land and labor.
5. The landless who work for a daily wage.

The interchanges of capital and labor follow complex patterns which I do not pretend to have fully unraveled. In the forty years since the ejidos were formed, the size of the individual parcels has decreased and the number of people without direct access to land has increased. Although the Chinantecs universally emphasize the fact and importance of individual land "ownership," only a few families actually do all the work on their own parcels themselves. Families are closely tied together by kinship and compadrazco. Tasks such as planting and harvesting, which require large numbers of people, are usually undertaken in groups. Two traditional systems of formal cooperation have survived to some extent. The first is more inclusive; a group will pool their resources and work together throughout the season, although the ultimate responsibility of each family for its own plot is maintained. The second is a work exchange system, in which one family will help another in its busy time and expect reciprocation.

These systems are slowly being replaced by cash; helpers will be paid by the day rather than being asked to wait for the favor to be returned or the harvest to be distributed. This increases the cash costs of production and reinforces the social differences that are based on economic distinctions. Grown sons of ejidatarios and other landless individuals often have secure sources of employment through long-term understandings.

The process of crop intensification on small units has diffused the problem of the less well connected campesinos who do not have direct access to land. There has usually been enough work for them to survive on a low standard of living. They do not work for a single plantation or a common boss, so they have never formed into a cohesive group or class. Every village has its poor. About half of the population is under 15 years old, so the problem of the landless would have become more serious in another ten years or so.

Since oxen are only used for three or four weeks a year on any one parcel, there is neither reason nor enough pasture for everyone to own a team. About a quarter of the families keep these animals, which implies relative wealth and status. Relatives and compadres get to use them at the most opportune times and at the best rates. This group shades up into the much smaller class of rich ejidatarios, who wield political power as well. Through their connections in the community and with the dominant mestizo clique, they have been able to acquire land and cattle. Herds of cattle and coffee plantations, even if they are not well managed or especially profitable, are concentrated in the hands of this group. These families have traditionally controlled the elected officer of the ejidos, and, as a group, the municipal chapter of the National Confederation of Campesinos (CNC) as well. This has given them access to the ruling political party (PRI) and to programs of the government, which they have distributed as patronage.

Sixteen percent of the zone that will be flooded is privately owned, some of it by minifundistas with less than ten hectares, but primarily by mestizo ranchers (Mexico, 27). Some of them are descendants of the employees of the old haciendas; others came into the area during the revolution. They form a community part, separated from the Indians by keenly felt class and racial barriers. They have their own formal association, and have much more in common with their counterparts in Tuxtepec than with their neighbors. Although their ranches are tiny by Mexican standards, they cause resentment by raising cattle on the kinds of high quality soils the Chinantecs use for intensive crop production. There are no figures showing how much ejidal land they rent illegally, but this does happen to some extent. They are an important political force in the Municipio.

THE FUTURE IN UXPANAPA

Although the Papaloapan Commission and other agencies have provided some services to the ejidos, the Chinantecs have not received their fair share of the benefits of regional development. The complex socio-economic structure in which they live has evolved over a long period of time. Piecemeal government programs have not been able to give the people an alternative basis on which to develop their resources for their own benefit. The official banks have provided agricultural credit to a few small groups, but their complex and inefficient procedures have not been able to provide a viable alternative to the paternalistic efficiency of the private lenders. In the late 1960s, the Ejidal Bank tried to reintroduce bananas into the area on a commercial scale to increase the income of the ejidatarios. Technical and marketing problems were never solved, the project collapsed, and the participants were left with a debt of over four million pesos (Bravo, Tamayo, Puya, 1972, p. 81).

In Uxpanapa, the Papaloapan Commission is attempting to build an entirely new economic and social system from scratch. The Chinantecs are being organized into production cooperatives large enough to use agricultural machinery and other inputs efficiently. It is hoped that this structure will allow them to overcome the restrictions imposed by the limited resources and exploitative economic system of their homeland. The Chinantecs have been farming in the rainy tropics for over a thousand years. It remains to be seen how they will be able to apply this experience on a large scale in Uxpanapa.

5 The Cerro de Oro Dam: Flood Control and Resettlement

THE PAPALOAPAN RIVER SYSTEM

The Cerro de Oro dam is but one element in a broadly conceived program of regional development in the basin of the Papaloapan River, which, with an average annual flow of 44 billion cubic meters, is the second largest river in the country in terms of volume. It drains a watershed of 46,500 square kilometers, or nearly 18,000 square miles. This area is about evenly divided between the lowland plains in the state of Veracruz and the highlands in the states of Oaxaca and Puebla.

The lower basin has been subject to occasional flooding for centuries. The capacity of the rivers to handle the run-off from the mountains began to deteriorate rapidly about 50 years ago. Population growth and rapid deforestation on the upper slopes accelerated the pace of the system's natural life cycle and made the floods more severe. Development in the lowlands meant that the economic losses were constantly increasing. According to a technical study made in 1943 (quoted in Rosales, 1977, p. 101):

> . . . before 1921, floods were infrequent, and only lasted three or four days. The agricultural lands benefited from the deposit of alluvial materials. Since then, both the frequency and duration of flooding have increased steadily. This has caused enormous losses to agriculture, and with time this fertile region may be rendered sterile by the leaching of the soil.

An especially disastrous flood in 1944, which covered 470 thousand hectares and caused millions of pesos worth of damage, brought national attention to the problem. The Papaloapan Commission, modeled on the Tennessee Valley Authority in the United States, was established in 1947 as a semiautonomous federal authority under the Ministry of

Hydraulic Resources. Its mission was to bring the river system under control, to coordinate the construction of roads and other infrastructure with a broad range of government services, and to bring the potential riches of the entire region into the national economy.

Erosion and Silt

Most of the precipitation in the Papaloapan basin is the result of orographic factors, as illustrated in Figure 5.1. The prevailing trade winds blow over thousands of miles of tropical ocean before reaching Mexico. This warm, moist air sweeps across the coastal plain in the summer, and is abruptly forced upward by the Sierra de Juarez, cooling as it rises. Most of the seasonal rain falls in a strip thirty to fifty miles wide along the windward slope, where both Ojitlan and Uxpanapa are located. As the air descends the leeward side of the principal ridge, it warms up and retains the remaining moisture. The transition between the tropical forest and the arid rain shadow is extraordinarily abrupt.

The Indians who live in the arid highlands are very poor, and have received very little benefit from the government since the Revolution. They have deforested large areas to survive. Less water is held by the exposed soil, run-off has increased, and erosion has become a very serious problem. The rivers carry the silt in suspension through swift-flowing gorges until they join the Papaloapan on the plain, where the water slows down and the river spreads out into leisurely, meandering curves. Sediment builds up, reducing the volume of water the bed can hold before it overflows its banks and spreads out over the flat lands on both sides. Other factors that influence the severity of a flood include the rainfall patterns in both the upland and lowland areas, the degree of saturation of the soil, and the flow of each of the major tributaries (Rosales, 1977). The annual average contribution of the principal rivers is illustrated in figure 5.2.

The Flood Control System

The keystone of the flood control program of the Papaloapan Commission is the construction of two large dams on important tributaries in the foothills just above the plain, at Cerro de Oro and Temazcal, Oaxaca. They will form a single reservoir of over 700 square kilometers with a maximum storage capacity of over 13 billion cubic meters - one of the largest artificial lakes in Latin America. Supplementary projects include several more modest dams on smaller streams, cuts to straighten meandering curves, and levees to increase the capacity of the main river.

Although the Cerro de Oro dam was considered by government engineers to be the single most important element in the system, the Temazcal site was less expensive to develop and that dam was built first, in the early 1950s. It controls the Tonto River, which drains 4,000

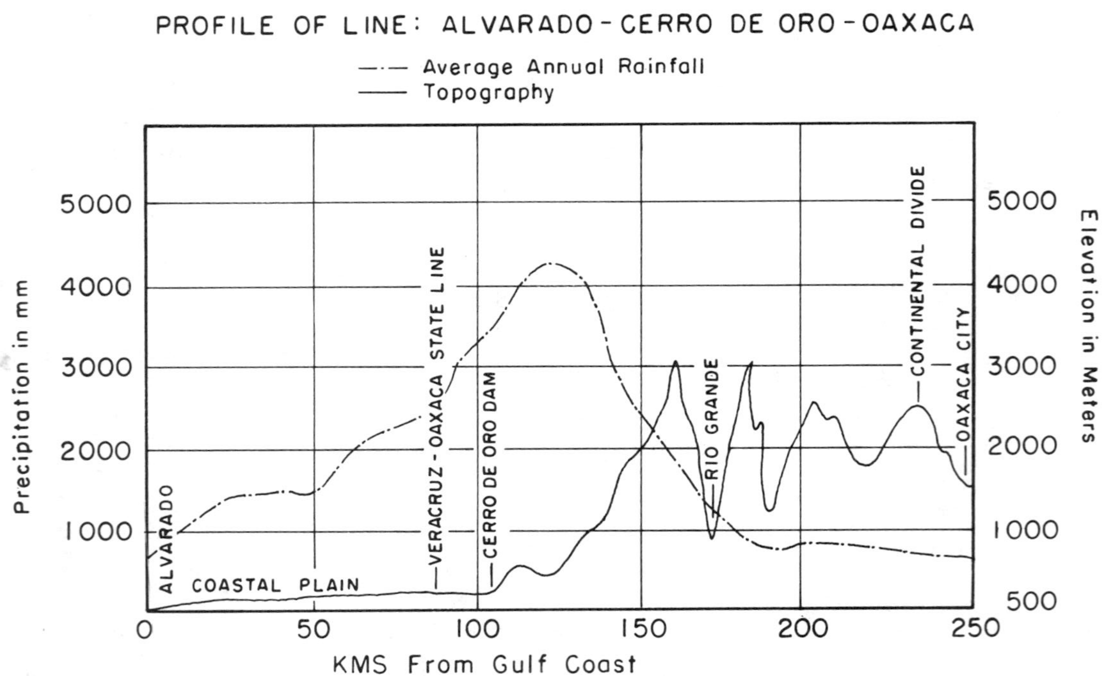

Fig. 5.1. Papaloapan basin: schematic relationship between topography and average annual rainfall.

Source: Adapted from Mexico, Secretaria de Recursos Hidraulicos, Comision del Papaloapan, Atlas Climatologico e Hidrologico de la Cuenca del Papaloapan, 1975, Plate 10.

(Not To Scale)

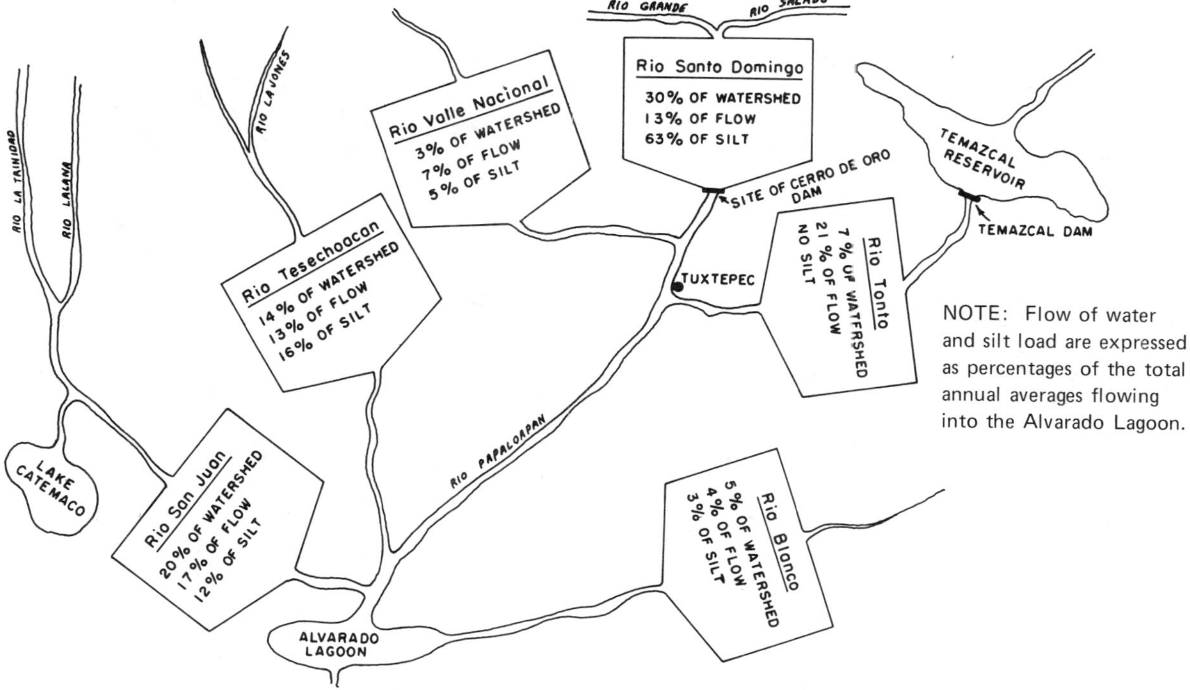

Fig. 5.2. The Papaloapan River system: water and silt contribution of each of the major tributaries.

Source: Mexico, Secretaria de Recursos Hidraulicos, Comision del Papaloapan, Atlas Climatologico e Hidrologico de la Cuenca del Papaloapan, 1975, Plates 22 and 23.

square kilometers on the rainy windward slope and contributed 21 percent of the water but only a negligible proportion of the silt to the total system. A hydroelectric plant with a capacity of 155,000 kilowatts was built and tied into a grid that serves both the industrial center of Cordoba and the Central Highlands. The Papaloapan was partially controlled and the construction of many new roads and other public works has led to significant development on the coastal plain in Veracruz (Poleman, 1964; Scherr, 1978).

The Cerro de Oro dam was not built for over thirty years, in part because national investment priorities favored other regions, especially in the Northwest, and in part because of strong resistance to the project in the state of Oaxaca. It is designed to control the Santo Domingo River, which drains 14,000 square kilometers including large areas of semiarid, overcultivated hillsides on the leeward slope of the principal ridge. This river contributes 13 percent of the water, but 63 percent of the silt to the total system. Siltration is rapidly reducing the volume of water the main bed can accommodate. In 1944, the capacity of the Papaloapan below Tuxtepec was 5,000 cubic meters per second. By 1972, in spite of the Temazcal dam, the levees, and other works, this figure had been reduced to 3,500 (Bravo, Tamayo, Puya, 1972, p. 25). Floods in 1958 and 1969 increased the pressure for the completion of the system. A year in which a hurricane or other factor produced unusually heavy rainfall was likely to see another severe flood, which would cause serious damage to the developed areas of the coastal plain.

The Cerro de Oro Dam

It was decided in 1972 to build the dam at Cerro de Oro and regulate the levels of its reservoir and that of the Temazcal by means of a connecting dike (Mexico, 38). The existing generating plant at Temazcal will be enlarged to a total generating capacity of over half a million kilowatts (Alatorre, Gapay, Morales: 1977). Combined with the construction of more levees and an extensive drainage system, the project will protect an area of 210,000 hectares from almost all danger of flooding. The probable useful life of the system is somewhat controversial. The Cerro de Oro reservoir will have a silt-holding capacity of 568 million cubic meters. If erosion in the watershed follows the patterns which have been established in recent decades, it will serve its functions for less than sixty years (Mexico, 28). More optimistic studies conclude that reforestation and other projects upriver will insure its usefulness for considerably longer (Bravo, Tamayo, Jimenex, 1972, p. 42). The Commission's existence is based on the concept that the entire watershed should be developed as a single natural unit. In its first 30 years, however, the great majority of both public and private investments were made in the lowland plain, where the returns were much higher than in the mountainous highlands. Between 1952 and 1976, only 50 million pesos were spent on soil conservation projects, about half in the upper basin (Mexico, 24).

The fundamental purpose of the dam was to protect the investments already made in the lower basin from the increasing probability of serious flooding. Nevertheless, it was necessary to include a broader spectrum of benefits in the project to gather enough political support for its construction. It was first proposed to create an irrigation district immediately downstream and relocate the Chinantecs there (Bravo, Tamayo, Jimenez, 1972; Mexico, 18). This would have required the enforcement of the statutory limit on individual land ownership, which was not well received by the local ranchers. When the technical studies were completed, it was concluded that seasonal irrigation would not be cost-effective.

In the early 1970s, the national agricultural crisis meant that all expensive projects in rural Mexico were justified by significant projected increases in agricultural production. It was finally decided to reorganize the area, which will be freed from the danger of flooding as a federal drainage district. This is a new administrative unit that theoretically gives the federal government the same kind of control over resource use it has long exercised over publicly-irrigated land. The present structure of land ownership of the district, almost all of which is in Veracruz, is shown in figure 5.3. Drainage and a variety of support programs are designed to increase the production of sugar cane and cattle, the predominant enterprises. Special extension and rural development programs will attempt to improve the precarious situation of the small farmers and ejidatarios (Mexico, 28). Soil conservation programs, roads, electricity, and other infrastructure will be built in the upper basin to encourage economic development and to reduce erosion (Mexico, 38).

Construction began in 1974, was interrupted in 1976, and began again in 1977. It should be completed sometime in the early 1980s. The dam will be 1,662 meters long, 59.9 meters high, 431 meters wide at the bottom, and ten meters wide at the crown. It will consist of a core of compacted earth secured with rock and gravel. Seismic activity has fractured the bedrock; concrete is being injected under pressure to secure the base. The process, called grouting, has proved more complicated than was anticipated. Three tunnels are being built for the completed system. The dam will flood 19,000 hectares of one of the most fertile areas of the Chinantecs' homeland. The Commission has been negotiating with the World Bank for loans to cover between 30 and 40 percent of the capital cost, estimated at seven billion pesos, or 30 million dollars, in 1977. It is an expensive project with a limited useful life that will eliminate the danger of flooding in the short run. It will affect both the natural and the social environment in ways that will be very difficult to predict or control.

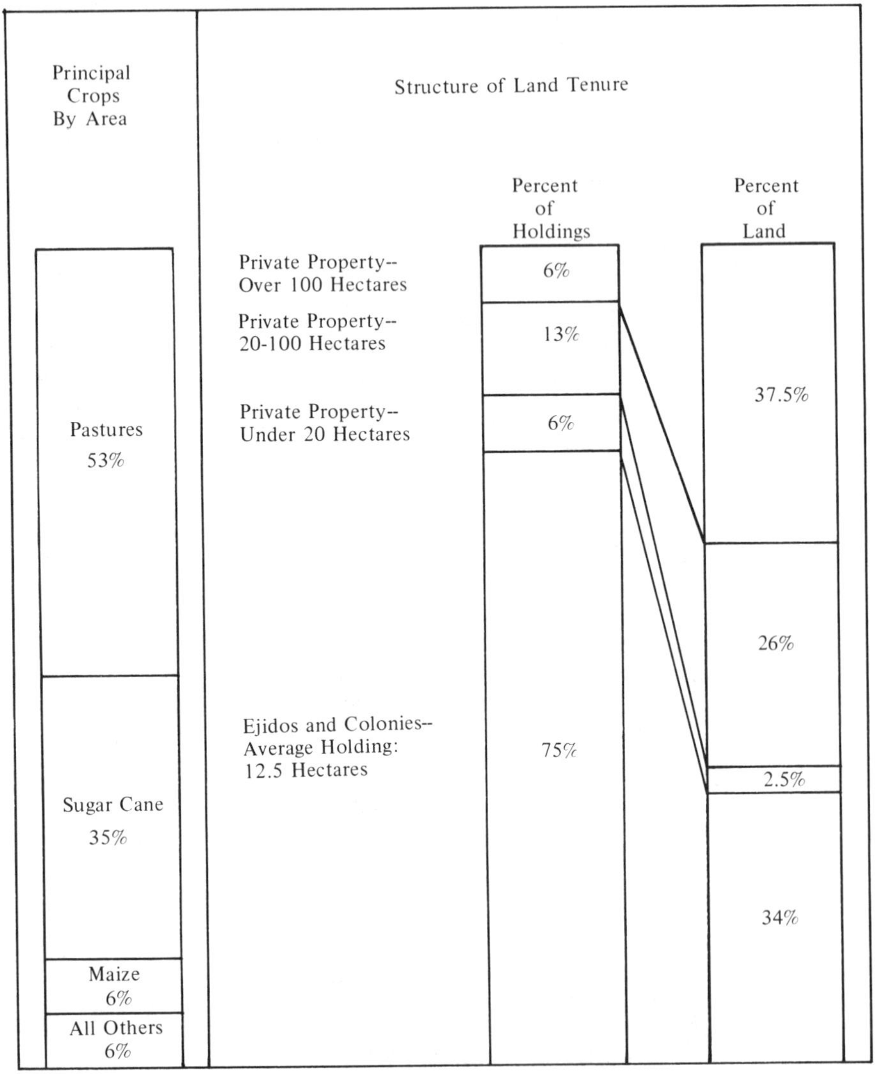

Fig. 5.3. The future Papaloapan drainage district: principal crops by area and structure of land tenure, 1976.

Source: Mexico, Secretaria de Recursos Hidraulicos, Subsecretaria de Planeacion, "Estudio Sobre el Proyecto para el Control del Rio Papaloapan" (mimeo, 1976), Tables 1:4-2,2:5-2.

THE RESETTLEMENT PROGRAM OF THE TEMAZCAL DAM

The Cerro de Oro project requires the relocation of 20,000 Chinantec Indians. The program has been strongly affected by the experience of the Temazcal, or President Aleman, dam that flooded 51,000 hectares in the valley of Tonto River and also displaced 20,000 people, most of whom were Mazatec Indians. It was the first major government activity in the area, and the Indians were much more isolated than the Chinantecs are today. The majority of them did not speak Spanish. The bulldozers that came to build the dam were the first pieces of machinery that the majority of them had ever seen (Benitez, 1970, pp. 15-22; Villa-Rojas, 1955, pp. 133-135).

The record of government-directed colonization programs in frontier areas, both in Mexico and in other parts of the tropical world, is quite poor. If new communities are to survive and prosper on the basis of viable agricultural enterprises, certain basic infrastructure such as roads, cleared land, and housing must be provided (Bassols, 1965; Lewis, 1964). On the other hand, it is very easy to fall into the trap of excessive paternalism by attempting to exercise too much control over the lives of the settlers, without giving them a personal stake in the project (Poleman, 1964, pp. 127-128). It is especially difficult to strike the right balance when dealing with Indian people with quite different values than those of the planners.

Today, in Uxpanapa, the government is attempting to use the resettlement program as the basis for a major agricultural development project set up to meet national goals. In the early 1950s, the policy environment was quite different, and the goals were much more modest. The Commission was obliged to compensate the Mazatecs with land in the other areas. The administration of the relocation process itself was delegated to the anthropologists of the National Indian Institute (INI). The program was designed to minimize the adverse effects of the forced move and to help the people adjust their traditional cultural patterns to their new conditions, and more generally to the modern world. On balance, the effects on the Indians were unfavorable, and the project created a precedent the Chinantecs were very unwilling to follow. On the other hand, it gave the Papaloapan Commission detailed experience it has tried to apply to the plan for Uxpanapa.

Although the planning and construction of the Temazcal dam took nearly five years, the resettlement program was poorly coordinated, beset with a variety of budgetary and other problems, and dragged out over nearly a decade. Three thousand families, a little over half of the original population of the lakebed, were moved into new areas under the auspices of the Papaloapan Commission. The rest took care of themselves, either by moving up into the mountains or by leaving the region. Alfonso Villa Tojas, an anthropologist who was in charge of the first stage of the process, was frustrated by the complexity of the task and the limited resources (Villa-Rojas, 1955, p. 37):

> One should not be surprised by the discrepancy which one almost always finds between the magnificent solidity of the physical constructions (of the Papaloapan Commission) and the fragility or the total destruction of the social organism which they are trying to benefit.

The Organization of the Move

INI was given less than a year before the dam was completed to inform the people of the program, enlist their cooperation, assess the value of their property, choose resettlement zones, clear land in the new areas, and provide housing and other basic infrastructure. They had to overcome the organized resistance of the mestizo ranchers and merchants, and of the Mazatec caciques whose power depended on the status quo. During the election campaign of 1952, both national and local politicians complicated the issue even further by promising that the dam would not, or could not, be built. The Tonto River was the axis of the Mazatecs' world, and many of them simply did not believe that any human agency could stop its flow.

Compensation for the value of the property that was flooded was very unequally distributed. Ejidatarios were eligible for indemnification for their homes, coffee plantations, fruit trees, and other immovable personal effects. Private owners were to be paid for their land and property. The Commission set a specific budget before the evaluations were made. The agrarian authorities set unit prices they felt would adequately cover the losses, but there was not enough money to cover the total. The mestizos and the well-connected Mazatecs were able to compete successfully for what was available. The cost of transportation, the new houses, and other servces in the resettlement zones was deducted from each account, so that only a few wealthy families actually received a cash balance. The process was slow - by 1964, only about three quarters of total had been paid (Vasquez, Villafranca, 1973).

The government committed itself to provide the affected population with parcels of land equivalent to what they had lost, or between ten and 15 hectares per family. It was originally intended to resettle the people in the area below the dam, that was to be served by an extensive irrigation system. A presidential decree was actually issued that expropriated 52,000 hectares for this purpose. Unfortunately, the first news of the project precipitated a speculation boom in this land, which had previously been isolated and nearly worthless. Investors from outside, including some employees of the commission itself, bought up large areas and lobbied successfully against its donation to the Mazatec people. Land prices rose all over the lower basin in expectation of the economic benefits of the dam and other new public works (Benitez, 1970, p. 18).

With a limited budget, the commission pieced together the necessary area. It could not relocate the population as a whole, or even the individual communities, in one place. It did acquire two portions of the original grant: an area in the immediate area of Temazcal, and the ex-hacienda of Los Naranjos. The State of Oaxaca, anxious not to lose so many of its citizens, donated the ex-hacienda of La Joya in an undeveloped area near Tuxtepec. To make up the balance, the commission bought two haciendas in an isolated frontier area to the south - Nuevo Ixcatlan (Yogopy) and Cihualtepec.

The project had suddenly become much more complicated than anticipated. Instead of being able to use the dam as a focus for a program of integrated development, the commission had to promise to clear land, build roads, and provide other basic services in widely scattered areas far from the center of its activities. Experience in the development of virgin tropical areas was very limited. Rapid surveys were undertaken from the air, but there was no basis on which to evaluate the results. Some of the Mazatecs were taken to visit the new zones, but only the small areas with relatively good soils near the rivers, that had already been settled, were accessible. Although the resources were poor, the officials of INI did the best they could with what was available (Villa-Rojas, 1955, p. 137).

> The extreme scarcity of good agricultural land in the Republic as a whole led us to think that if the zones . . . were at least above the national average there was no reason to reject them. To hold out for first class soils appropriate for mechanization would have been overly ambitious and utopian.

The Mazatecs, who had been suspicious from the first about the intentions of the outsiders, found their worst fears confirmed. Their ancestral land and goods were being expropriated without proper compensation, the integrity of their communities was being destroyed, and their people were being scattered into hostile jungle areas among a mestizo population that exploited and looked down on them. They turned into their traditions for weapons with which to fight back.

In 1954 the dam engineers closed the gates of the completed dam to test the system. As the water rose over the fields, the elders and religious leaders held ceremonies, asking the spirits to sweep away the ugly wall that threatened their lands. The next day, their test completed, the engineers opened the gates and the flood subsided.

By working hard against an impossible timetable, INI was able to move 860 families into the nearest of the zones before the reservoir began to fill. Nevertheless, large numbers of people were still in the lakebed, in part because the new areas had not been prepared, and in part because they had refused to leave. They retreated onto the hills, which quickly became islands. The churches were packed with Mazatecs holding candles, imploring the saints for aid. The commision sent out teams in launches to rescue the refugees (Benitez, 1970, pp. 26-27):

> An old Mazatec owned a series of three huts running up the side of a hill. When the water reached the first cabin, he moved his animals

and belongings into the second and refused any kind of help. A few days later, when the water reached the second cabin, he moved up into the third, declaring that he would stay there "even if the water comes up to my neck." The anthropologists respected his decision, but the next morning, when the launches were passing nearby, they saw the stubborn old man making signals with a cloth. They approached, and he accepted his rescue. "It wasn't the water that drove me out, but the snakes. My son was bitten this morning and is in a bad way. There is no way we can get to the culebrero." (A culebrero is a traditional medicine man who specializes in snake bites.)

The New Communities

This section is based on interview and on the following sources: Avila, 1977; Ballesteros, 1970; Benitez, 1970; Cisneros, 1973; McMahon, 1973; Nelson, 1973; Poleman, 1964; and Villa Rojas, 1955. The five settlement zones were widely scattered and encountered distinct problems. A rapid survey of their history will illustrate that range of the Papaloapan Commission's experience with tropical colonization.

Temazcal - A Social Experiment

Of the new colonies, only the area immediately downstream of the dam was fully planned and at least partially established before the flood. Nine new villages were built to accomodate 860 families, both Mazatec ejidatarios and mestizos. Although no systematic attempt was made to change or improve the traditional agricultural technology, it was hoped that the paved access road and closer contact with the outside would stimulate development. The Indians were given the same kind of housing and services as the gente de razon in separate neighborhoods. Although this segregationist policy seems a little naive now, the belief that if people could live like equals they would begin to behave like equals was common in rural development theory in the 1950s. After a short period of disorientation, the merchants and caciques quickly reestablished their economic and political dominance. Some of the Mazatecs have joined cooperative fishing ventures. Although an expected influx of tourists never materialized, the general prosperity of the Tierra Blanca region has meant that the people who remain live at least as well as the average smallholders in the area.

Los Naranjos - Excessive Paternalism

Los Naranjos, intended as a pilot project for a 160,000 hectare irrigation district that proved impractical, was divided up into colonies by the commission in 1953. In two of these, "Oaxaca" and "Nuevo San Jose," a few smallholders from the lakebed were granted ten hectare plots of land. Irrigation water was supplied to a small area for two years, but this aspect of the program was abandoned as too costly. Credit was provided on a closely supervised basis for the rain-fed production of various crops, particularly rice. The costs of an extensive infrastructure were charged to the campesinos. Except for one colony that was allowed to work independently, the settlers were forced to accept a high degree of control from the commission's technicians. Machinery was provided on a custom basis, and field supervisors directed every step of cultivation. There had never been any kind of experimentation in this kind of agriculture in this climate, and there were a series of expensive failures. The project itself was not set up as an experiment from which to learn, and the colonists had to accept all of the risks.

In 1957, when the commission's budget for agricultural operations was reduced, the colonists received orders to pay the total value of the land, the infrastructure, and the past losses within four years. No bank was prepared to refinance under these conditions, and many of the people were forced to leave.

La Joya - Insufficient Infrastructure

In 1954, there were already over a thousand families living in the La Joya region, occupying most of the good land. When the gates of the dam were closed, the water rose more rapidly than had been anticipated, and it was necessary to move in 450 families before adequate preparations were made. A road was built from Tuxtepec. When the land was cleared, it was discovered that the quality of the soil had been seriously overestimated. The people were in an acute state of shock; they had been moved hurriedly from their ancestral lands into an area they did not like. They felt that they had been deceived about its quality, and that they were not receiving the treatment they deserved. The commission tried to introduce perennial crops including coffee, natural rubber, citrus fruit, and mangos. The settlers did not have the land to grow enough maize to feed their families. They were unwilling to give over any of it to crops that would only bear in a distant and uncertain future. This lack of confidence was compounded when an experimental rubber planting died of disease. People from different areas were resettled together; they separated into mutually-antagonistic groups.

When their budget was reduced in 1960, the commission and INI abandoned all but the most essential services. They stopped maintaining the road, which was completely impassable within a few months. River transportation was slow and expensive; the people were reduced even further to a level of basic subsistence. One of the groups had a strong leader who was supported by INI. Without any kind of government assistance, he organized the migration of over 30 percent of the colonists to Cihualtepec. The others stayed on, feeling hopelessly abandoned in the jungle. The road was reopened in the late 1960s, a new sugar mill brought land into cultivation, and La Joya finally began to participate in the development of the region.

Nuevo Ixcatlan and Cihaultepec - The Colonists Lose Their Land

The last of the new communities were located in a heavily forested area in the valleys of the Lalana and Trinidad Rivers. American companies had taken out precious timber at the beginning of the century, but settlement was limited to narrow bands along the rivers. Twelve thousand Mazatec, Mixe, and Chinantec Indians were relocated in the area between 1954 and 1967. Most of them came from the upper end of the lakebed. The reservoir took several years to fill, so the move was organized under less pressure than in the other cases. The natural environment and the social issues involved are quite closely analogous to those faced in Uxpanapa today. The experience provides a good example of the kinds of problems the Papaloapan Commission is trying to avoid through centralized management and collective organization.

The forest was cleared by the commission and four sawmills were set up to commercialize the timber. Neither the technical control nor the markets were available for its efficient utilization; approximately 75 percent of the potentially valuable wood was wasted. The people were paid a daily wage to build their own houses, which provided an income until the first harvest.

The land was divided into ten hectare parcels. Except in small areas of alluvial soils, the decline in fertility was very rapid; average yields decreased by 50 to 60 percent by the third year after clearing. From then on, only three to four hectares were planted in each parcel and the rest was left in fallow. This was barely sufficient to grow enough to eat. During the same period and in the same zone, colonies organized by the agrarian authorities (DAAC) were granted 50 hectares per family. Private cattle ranches had to be at least 500 hectares to attract buyers. The commission was obliged to give the Indians an area equivalent to what they had lost, but it was unreasonable to expect that they would be able to overcome their traditional poverty on such a meager resource base.

In recognition of the obvious advantages of a high-value crop with high labor requirements, a credit program for tobacco was introduced. It requires constant care, specially designed drying sheds, and careful timing. The Mazatecs had never grown it before, and the technical assistance was poorly coordinated. There were serious problems with pests. The commission promised to build a drying shed, but never finished it. The leaves did not dry well, the Indians did not know how to separate the harvest into different grades, and the buyers paid the lowest possible price for the whole crop.

The commission also provided credit for rice the people had grown in their homeland. Vital inputs arrived late, payments were delayed, and debts built up. When the colonists tried to sell the crop to the local merchants to meet pressing needs, the resident administrator blockaded the road to embargo the grain until the credit balance was paid. Several other officially sponsored agricultural programs ran into similar problems.

The mestizo merchants and ranchers quickly reestablished the dominant position in the local economy they had enjoyed in the lakebed. The anthropologists from INI undertook a number of programs that attempted to break down racially-based roles and systems of exploitation. They believed that the problems were built into the culture itself, and promoted changes in the way the Indians lived as the basis for their liberation. Bilingual Mazatec "promoters" or "change agents" were trained to serve as a bridge between the traditional ways and the modern institutions. Rather than attempt to transform the culture in one jump, they tried to introduce key elements into the indigenious lifestyle they hoped would lead logically to a new world view and new patterns of behavior.

The unsuccessful result of one such program is the first thing a visitor notices upon entering one of the these towns. At one end of each lot is a thatched house; at the other is a concrete frame overgrown with weeds and vines. The Mazatecs have built their houses according to the same basic design for centuries. The anthropologists felt that the floors of compacted earth were unsanitary. They also believed that having a dozen members of an extended family living in one room led to promiscuity and prevented the children from defining themselves as individuals with productive roles. As the colonists arrived, INI built skeletons for improved housing that consisted of a concrete floor and six posts. They hoped that the people would go on from there and modify their traditional patterns to accommodate these and other modern elements.

The Indians ignored the idea, built their houses in the traditional manner, and used the concrete platforms to dry maize and barbasco. Interpretations of this phenomenon vary, and pose the question of what aspects of the people's lives were the legitimate concern of the government. The mestizos laugh about how the paisanos prefer to live like pigs. The anthropologists have discovered that the Mazatec bury their babies' umbilical cords in the floor, and feel that some religious value is involved. The promoters are acutely embarrassed by the "backwardness" of their people. The Indians themselves say that the concrete is cold to sleep on, and is actually more difficult to keep clean than well-compacted dirt, which is swept and sprinkled every day. In any case, the government spent a fair sum of money following an intellectual concept rather than respecting the priorities of the Mazatecs themselves.

INI set up a store in Nuevo Ixcatlan to sell agricultural inputs, foodstuffs, and other necessities at prices below those charged by the private merchants, but it could not compete with the private stores. INI tried to enforce prohibition, and did not sell beer. The private merchants kept large stocks of liquor under the counter and were always willing to make some loans to their compadres until the next harvest. The official store depended on a cash flow of freely disposable income that was simply not there. By purchasing the crops in advance, the mestizos were able to control an economy that operated almost entirely on credit. The store belonged to the "the government," which had torn the settlers out of their homeland and turned them into strangers in a difficult environment. Their experience with official programs was overwhelmingly unfavorable, and they felt patronized by the interest in "improving" their culture. Few people paid their bills and the store went bankrupt.

As the area developed, cattle operations on a fairly large scale were increasingly capitalized and became more profitable. At the same time, the fertility of the soil in the ten hectare grants fell below the point where they could support a family. Unless it was fertilized intensively pasture was the only economic use of the land. The Mazatec families did not have the resources to buy either cattle or the inputs necessary to maintain crop yields. They began to rent or sell their parcels to the ranchers, who were often the same merchants to whom they owed money. The ranchers owned large tracts of uncleared land that they turned around and rented to the colonists. The campesinos would clear plots, plant their crops, and reap the benefits of the first flush of fertility. After a few years, these areas would in turn be planted with pasture. The cycle continued as long as suitable and accessible land was available.

This sequence - the clearing of virgin forest, the cultivation of crops for a few years to take advantage of residual fertility, and the establishment of permanent pasture - has been repeated countless times throughout the tropical world. It is the path of least resistance, and has undermined the goal of the project, which was to provide the people displaced by the dam with a permanent source of income on land of their own.

The municipal president of Cihualtepec estimated in 1977 that over 70 percent of the original colonists had no land at all. Their alternatives were so limited that very few of them had left. They survived by taking jobs as day laborers on the ranches where there was work, and by eking very small yields out of subsistence plots. The government tried to organize a collective credit system in the 1970s, but the basis for a program of this kind had already been eroded away. Only five small groups in peripheral areas had been organized by 1977. The gente de razon had acquired control over almost all of the profit-generating resources in the area; the land, the cattle, the transportation system, credit, and marketing.

The government's investment in roads and other infrastructure was successful in opening a frontier area to economic activity. Between 1954 and the late 1960s, the population rose from 3,000 to 18,000 and the area in production rose from 25,000 to 150,000 hectares. The major paved road from Tuxtepec to Palomares, which runs through both zones and was completed in 1976, will certainly encourage increasing economic activity in the future. It could be argued that ranching is really the most appropriate use for the land in this ecological zone, but the social purpose of the project has failed.

THE ANTI-CERRO DE ORO MOVEMENT

Unlike the Mazatecs 25 years before, the Chinantecs of the Ojitlan area were very aware of the impact the construction of the Cerro de Oro dam would be likely to have on their lives. The Temazcal reservoir is clearly visible from the main square of the town. Ixcatlan, which survived the previous project as a lakeside town with a greatly reduced economic base, has been in many ways a sister community for centuries. The two peoples do not understand each other's language and seldom inter-marry; they have lived very closely together. Some Chinantecs were affected by the previous project, and some Mazatecs are being affected by this one. The story of the resettlement projects was well known.

The Regional Conflict

The Cerro de Oro project is an integral part of the flood control plan first formulated in 1948. Since then, almost every one of the Papaloapan Commission's administrators had made new studies and sought in vain to get national support for its construction. When the process was reinitiated in 1972, a coalition of interests opposing the dam had every reason to believe that an organized response would again lead to its indefinite postponement.

Opposition to the project had been a key point in every local election campaign for almost a generation. We have seen that poorly distributed development had taken place during this period. The groups that benefited from the process wielded economic and political power; and were allied with their counterparts in Tuxtepec and Oaxaca. The Chinantecs had no interest in being forced into a new situation over which they had no control. Their experience with the government consisted of a long series of unfulfilled promises and projects that had only benefited a minority.

The line between the states of Oaxaca and Veracruz marks the transition between the foothills of the mountains and the coastal plain along much of its length. Both Temazcal and Cerro de Oro are less than 20 kilometers from the boundary. The first dam was designed to prevent flooding in the lowlands and most of the commission's subsequent investments were made in Veracruz. Oaxaca is a poor, mountainous state where the Indian cultures predominate. One of the few areas that has produced a significant agricultural surplus is the small tropical lowland zone around Tuxtepec, where the prosperous mestizo ranchers share strong family ties and are an important political force. The federal government proposed to flood another fertile valley in Oaxaca to protect the coastal plain. Although there would be some local benefits, this sacrifice would fundamentally favor the same interests across the state line who had profited from the Temazcal project.

The three groups - the Indians who feared that they would suffer the same fate as the Mazatecs, the commercial elite whose income and position depended on the status quo, and the land-owning elite who feared that their interests would be sacrificed to benefit another region - formed a powerful coalition. President Echeverria visited Tlacotalpan and Tuxtepec in March, 1972, as part of his concentration on the impoverished South during the "Year of Juarez." When he announced that the dam would finally be built, he was faced with organized and broadly-based opposition, which is unusual in Mexico and was played up in the Mexico City press. The National Confederation of Campesinos (CNC) organized an anti-dam committee, supported by such diverse groups as the Cattle Associations, the Association of Private Landholders, the Cane Growers Association of the local sugar mill, the Chamber of Commerce, and the Lion's Club (Lucero, Avila, 1974).

The New Emphasis on Oaxaca

The former governor of Oaxaca and minister of education, Victor Bravo, is a member of Tuxtepec's land-owning and commercial elite. Jorge Tamayo, a distinguished Oaxacan geographer, was soon named head of the Papaloapan Commission and the Tuxtepec Paper Factory, the area's most important industry. Echeverria appointed both of them to a Special Study Commission which recommended that federal investment priorities in the Papaloapan Basin be reoriented toward the Oaxacan portion to break down the local objections to the project (Bravo, Tamayo, Paya, 1972). The executive decree that authorized the construction of the Cerro de Oro dam included programs for new roads, electricity, water systems, schools, hospitals, church renovations, and forestry and agricultural projects in the upper basin, including the area to be flooded (Appendix). A large delegation of Chinantecs was invited to the presidential palace in Mexico City, where Echeverria personally assured them that they would be paid generous indemnification, and would receive free land and comprehensive services in whatever areas they themselves chose for their resettlement. The whole effort was an expensive, but effective, solution to the political problem. It was also consistent with the general policies of Echeverria's administration, which was attempting to decentralize federal investments and to promote a more equitable distribution of income, particularly in depressed rural areas such as Oaxaca.

Political and Emotional Turmoil

As the dam project came closer and closer to becoming a reality, the people of the Santo Domingo valley went through a period of severe political and emotional turmoil. The political and economic structure was called into doubt, and the culture that had developed in the area over a millenium underwent a serious shock. The local chapters of the official national party (PRI) and the National Confederation of Campesinos (CNC) had backed the wrong horse. They had long enjoyed unopposed power, and had rotated the various offices among the members of a small elite. In the municipal elections of 1972, they were faced with a strong opposition running on a pro-dam platform under the auspices of a minor party (PARM). This group consisted both of people who believed that they would get a better deal by supporting the new policy, and of the political "outs" who saw a chance to gain power at a moment when a large influx of money from the outside was obviously on the way. The PRI didn't lose many elections and they didn't lose this one. Convinced that the ballot boxes had been stuffed, the PARM candidates occupied the town hall. Their claim was recognized by the state with remarkable speed, and they took office.

The CNC was founded under Cardenas' administration, and represents the overwhelming majority of the ejidatarios in Mexico. Along with the industrial labor and civil service sectors, it is one of the three constituent elements of the PRI party. It does not represent the campesinos as an independent bargaining unit, but serves as the medium through which they are integrated into the monolithic structure of national political power. In Ojitlan, an overwhelmingly ejidal municipio, the local chapter amalgamated the interests of the mestizo and Chinantec elites. They now found their position challenged by a small, much more radically-oriented campesino organization founded only ten years before, the Union of Independent Campesinos (CCI) (Gomez, 1970). The serious divisions that have developed in the ejidos as a result of the dam project have been polarized into a conflict between these two groups.

An emotional religious movement swept through the Chinantec communities in 1972. The Virgin of Guadalupe, Mexico's patron, appeared in a cave to a campesino who was walking by himself in the hills (Mexico, 26):

> My son, do not be afraid! I am the Virgin of Guadalupe, Mother of all the Mexicans and Queen of America. I have come to protect you all against the Cerro de Oro project, because I do not want you to suffer as my children the Maztecs have suffered.

The story follows the original sixteenth century miracle very closely. The Virgin asked the man to bring the municipal president to speak with her, but he refused to come. In spite of a long standing taboo against entering caves, hundreds of pilgrims made their way to the site, to light candles and offer their prayers. A few months later, an angel who identified himself as God's personal engineer came to assure them that the dam was a technical impossibility (Barabas, Bartolome, 1973, p. 6).

The government was faced with what was very obviously a difficult situation. A path through the different factions had to be chosen with great care. The project received the unqualified support of the governor of Oaxaca and the minister of hydraulic resources. INI was largely excluded from the process because of a variety of conflicts with the Papaloapan Commission that had arisen in the course of the

Temazcal resettlement program. A group of social scientists was hired by a specially-created Human Relations Commission to bridge the gap between the government and the Chinantecs.

1972-1973 was a very confused and difficult time. The future was uncertain, the people were divided into a number of antagonistic groups, and emotions were running very high. Many of the social scientists who were in constant contact with the campesinos were shocked both by the degree of local exploitation and by what they saw as the heavy hand of the government. Many of their reports were critical both of the dam project itself and of the means the federal authorities were using to impose it on the Indian people. One of the very strong statements was written by a pair of anthropologists and published in Europe (Barabas, Bartolome, 1973, p. 3):

> The aim of this report is to call attention to a program of ethnocide which is being applied to ethnic minorities in Mexico as part of a program of capitalist development being carried out... through regional development agencies. The aim is the incorporation of the Indians into the national capitalist system of production and consumption (through) the elimination of their economic semi-independence and cultural identity ... to incorporate the Indian into a rural proletariat.

Other reports were more accurate, but they did try to open up a dialogue in the cities of Oaxaca and Mexico City over the appropriate balance between national development goals and the rights of the Indian minorities. The Papaloapan Commission became very defensive, and increasingly centralized the administration of the project into the hands of its own technical staff. Gonzalo Aguirre Beltran, an anthropologist and distinguished elder statesman of Indian affairs in Mexico, was outraged by what he saw as a rising tide of irresponsible radicalism (Aguirre Beltran, 1957).

The reaction is understandable given the background of the people involved, but it marks the beginning of a fortress mentality that has characterized the program. Although a great deal of information on the economic, social, and cultural structure of the Chinantecs was gathered in this period, there has been almost no input by anyone trained in the social sciences since. Almost all of the energy has been put into the physical construction of the visible elements of the project. The program is set up to give a very large number of real benefits to the Indians. These benefits are conceived, designed, and administered by technically-trained people organized in a rigid hierarchy. As we shall see, the lack of emphasis on the more subtle social processes, and the lack of consultation with the Chinantecs themselves, has resulted in increasingly serious problems.

THE SELECTION OF UXPANAPA

The public works program for the upper basin effectively divided the broad coalition of regional interests that had opposed the dam. Nevertheless, various groups and factions of Chinantecs continued to send petitions to Mexico City, demanding that their rights be respected. They were personally assured by Echeverria that each family would be provided with twenty hectares of good land served by an extensive infrastructure. A resettlement committee was organized in Ojitlan in 1973 to coordinate the program. The problems involved in finding 80,000 hectares suitable for agriculture not already in use were even more complex than they had been when the Papaloapan Commission faced exactly the same problem in 1953. Most of the frontier had been settled. Some of the area immediately surrounding the new reservoir was available. Although it formed part of the traditional homeland, most of the good agricultural land was to be flooded. Only a small portion of the affected population would be able to make a living there with cattle, tree crops, and fishing cooperatives.

Delegations of Chinantecs were chosen to visit potential sites in other areas. Several zones in Oaxaca on the eastern slope of the mountains were considered, particularly an area near the old resettlement zone of Cihualtepec. These lands were only appropriate for extensive ranching, and the people expressed a very strong preference for continuing as farmers. The search moved outside of the commission's jurisdiction, to the wild and isolated mountains in the municipio of Santa Maria Chimalapa in the Isthmus of Tehuantepec. This area is the home of 70 Zoque families, who have maintained continuous communal ownership of an area of 400,000 hectares since before the Spanish conquest. Most of the land is too steep even for cattle, but the exploratory investigations led to what looked like a very good area immediately across the Veracruz state line to the north - Uxpanapa.

The government had intended to respect the Chinantecs' desire to stay in Oaxaca and use the resettlement project as an opportunity to develop another area of the state. Uxpanapa, however, was by far the best alternative they found. The climate is similar to that of Ojitlan. The area was almost entirely covered with virgin tropical rain forest. It had been divided into ejidos in the 1960s, but very few campesinos had actually taken possession, so the land was already under federal jurisdiction. The population living in an area of 260,000 hectares was under 2,000 at the time. There were no access roads, but the road from Tuxtepec to Palomares, directly linking Ojitlan and the entrance to Uxpanapa, was already under construction.

The commission could not justify the enormous investments involved in developing such an isolated virgin area only to give several thousand families a place to live. They formulated plans to open up Uxpanapa for intensive agricultural production to meet the increasing needs of the nation and to serve as a model for developing the over six million hectares of tropical rain forest in the Southeast. After several months of study, the area was officially designated as a drainage district for resettlement under the administration of the Papaloapan Commission (Mexico, 39).

Of the 51 ejidos in the future Cerro de Oro reservoir, 36 will be completely flooded and 14 will be compensated for part of their land. Although Uxpanapa was strongly favored, the Chinantecs were presented with three other options. Part of the same Los Naranjos district, unsuccessfully promoted in the 1950s, was to be provided with irrigation, housing, and other infrastructures. A small number of vacancies in the ejidos on the periphery of the reservoir were available. The region near Cihaultepec was also offered, but given the experience of the earlier colony and the inferior quality of the soils, it is not surprising that it was not accepted by anyone. Each family was offered ten irrigated hectares or twenty hectares in the rain-fed zones. The commission provided transportation for all of the household heads to visit the areas and make their own decisions.

Some were taken to Uxpanapa by helicopter. The vegetation was so dense that in many cases a man was lowered on a rope to cut out a landing space with his machete. The majority were taken up the rivers in launches.

Camps were set up at the existing settlement of Hermanos Cedillo and on the Uxpanapa river. Groups would spend a week or more exploring, hunting, and evaluating the land as best they could. Many men were hired by the commission to work on the clearing and the construction. Both the topography and the climate of the zone are highly variable, and the dense forest cover allowed only a superficial analysis (Mexico, 25).

Meetings were held in each of the ejidos, and the resettlement zones were chosen by majority vote. Almost all of them were bitterly divided. The members of the CNC were still basically opposed to the project. They argued that Uxpanapa was far away in another state and completely underdeveloped. The people would be totally at the mercy of the government, which they did not trust. It is not easy to generalize about the social composition of this faction, but it tended to include the people who had been relatively powerful under the old system and their dependent groups, as well as conservative Indians who did not want the Chinantec traditions and customs to be lost. If they could not stop the project, they insisted on the Los Naranjos zone because it was closer to civilization. Given the earlier history of that project, its future prosperity seems doubtful, but they felt that it would give the settlers greater flexibility. For a variety of reasons which will be discussed at length, the commission insisted that Uxpanapa be organized into collective ejidos, or cooperative production units. Although each family was given rights to 20 hectares, the land in each ejido was in fact combined into a single unit. Almost everybody would have preferred individual assignment of parcels, and this became a key point in the opposition.

The CCI has worked very closely with the commission. The national organization is strongly committed to the concept of collective ejidos.

This group organized the people who liked Uxpanapa and who felt that by supporting the government program they would be able to insist on all of the services they had been promised.

It should be remembered that the majority of the population was illiterate and that many could not speak Spanish. The commission made a tremendous effort to observe democratic procedures, but the politics of the situation were dominated by a limited number of leaders in the various factions. Most of the Chinantecs accepted the reality of government power and sought to make the best of it.

When the votes were taken the results were as follows (Mexico, 25):

Resettlement Zone	Number of Ejidos	Number of Families
Uxpanapa	28	2,796
Los Naranjos	9	946
Peripheral Area	14	N.D.
Choapan (near Cihualtepec)	0	0

These figures include all men over 16 years old at the time of the special census, whether or not they had previously held ejidal rights. The future status of the minorities in each ejido is still unclear. Many of them will eventually accept what they are being offered. At least one group had independently organized a move into an area in the state of Chiapas. Some individuals have taken their indemnification and have moved into Tuxtepec or other areas.

The agrarian authorities assigned the people into the vacant places in the existing ejidos in Uxpanapa. The move was organized in stages, to allow time for the construction of the new towns and other facilities in the resettlement zones. The campesinos retained the right to continue to live in the lakebed and cultivate their lands until they were actually flooded. Private landowners were indemnified in cash after all the ejidatarios were paid. Most of the mestizo merchants lived in the town of Ojitlan, which will not be flooded. The comprehensive plan for Uxpanapa discourages them from moving into the new areas and reestablishing their control over the economy of the Chinantecs.

INDEMNIFICATION

Both the Chinantecs and the commission were very aware of the inequities and problems that accompanied the indemnification process in the Temazcal program. The government undertook a careful and expensive program to insure that the valuation would be complete and fair, even if it meant paying more than the market value of many items. One hundred bilingual inspectors were hired. They worked in the ejidos in pairs. One would write down the value of each house, fruit tree, coffee bush, meter of wire, and so forth, while the other would mark each object with red paint. A chief inspector would later verify the evaluations, and any discrepancy was carefully checked. An individual inventory was drawn up for each campesino, which he would sign or mark in the presence of witnesses. Communal property of the ejidos, such as schools, basketball courts, and assembly buildings, were evaluated in the presence of the local authorities. Along with the indemnification for the land, which was valued with standard tables according to quality, these funds were credited to the ejidos' accounts in a national financial agency (FONAFE). This money was to be drawn out for capital improvements and community services in the resettlement zones once certain criteria were met. A check for the amount of his personal property was given to each campesino.

Twelve of the ejidos were paid in 1975, three more in 1977, and the balance in 1978. (The average price paid for land was 3,600 pesos per hectare, or between 65 and 115 dollars per acre depending on whether one uses the pre- or post-1976 exchange rate; the average payment for personal property was 33,000 pesos, or between 1500 and 2600 dollars.) The payments were distributed as follows (Mexico, 14):

Category	Amount (pesos)
Land	87,056,500
Common Ejidal Property	2,954,300
School Property	5,305,600
Personal Property	133,062,600
Total	228,378,000

These figures have been adjusted somewhat since the original evaluations were made. The indemnification has reflected the distribution of wealth in the community, and was skewed to the high end. Richer campesinos have had their assets liquidated for them, and have bought trucks or gone into business. As in the Temazcal program individuals with extensive coffee plantations have received the highest payments. All bushes were valued at a flat 80 pesos, regardless of their quality.

Free transportation to the resettlement zones was provided for the people and for their movable personal property such as household goods and domestic animals. It is remarkable how little most of them owned. A typical inventory consisted of two sheets of corrugated zinc, two dozen boards, a table, three chairs, two cots, a picture of the Virgin, four pigs, two turkeys, twenty chickens, a dog, and three bundles of clothes and cooking utensils. Since they are only allowed one trip and many of them are maintaining two households during the transitional period, many families have transported their goods at their own expense.

The employees of the commission who have worked in Ojitlan deserve credit for having administered a fair and orderly procedure. They moved into a chaotic situation and earned the respect of most of the Chinantecs. Their daily work consisted of an endless series of petty details; the organization of trips to Uxpanapa, a pig left behind, rumors that the indemnification would never be paid, and conflicts over wages. At least in its early stages, the program had an ample budget. The delay in the construction of the dam gave them much more time than was available in Temazcal. This stage of the project was pretty well over by 1976; the center of gravity had shifted to Uxpanapa itself.

A Chinantec woman.

Although each family farms an individual plot which averages between one and five hectares, groups of relatives and friends cooperate for land preparation, planting, and harvesting. The minority with the resources to support a team of oxen rent them out to their neighbors.

A few settlers have built traditional structures adjoining their "official" houses. Others have sent relatives to build the house and receive the subsidy, but continue to live in Ojitlan.

The spontaneous settlers cleared small plots in the tropical forest by hand to meet the needs of their families.

Upland rice was also grown using traditional slash-and-burn methods.

One of 12 new towns that are being built in Uxpanapa for the Chinantec Indians who are being displaced by the Cerro de Oro dam.

The Papaloapan Commission is building a civic center in each of the 12 new towns, consisting of a primary school, a meeting hall, a health clinic, and a post office.

Combines have not been able to harvest upland rice effectively under the rainy conditions in Uxpanapa. The chief agronomist tries to adjust the machine to cut lodged grain.

Twenty hectares were planted per ejidatario, much more than they could possibly harvest by hand when the available combines were incapable of completing the job. Hired workers from outside the zone demanded higher pay than the value of the harvest could support.

Cattle enterprises are being introduced. The herds, composed of Cebu - - Holstein -- Brown Swiss crosses, will produce both milk and meat.

6 Uxpanapa: The New Land

THE ISTHMUS OF TEHUANTEPEC

Uxpanapa is located squarely in the middle of the Isthmus of Tehuantepec, a 125-mile wide neck of land that marks the natural transition between North and Central America. The great spine of mountains that runs the length of the continent breaks into lower hills and narrow passes which, since the days of Cortes, have excited interest as a means of communication between the Atlantic and Pacific Oceans. Its northern half lies in the state of Veracruz, and is a rolling tropical plain of dense forests and extensive swamps. The Chimalapa mountains in the center, whose highest ridges rise 1,000 meters above sea level, are so wild and inhospitable that no one has ever challenged the rights of native Zoque Indians since colonial times. On the leeward side, the lowlands of the Pacific plain in Oaxaca are arid, covered with thorny bushes, and buffeted by winds of such force that they sometimes blow trailer trucks off the Panamerican highway. The towns of Tehuantepec and Juchitan are the cultural centers of the enterprising Zapotecs. A series of lagoons along the coast are the home of the Huaves, one of the poorest and most isolated native peoples in the countries. The Mixes live in the mountains to the west.

Until recently, the Isthmus was more of a barrier than a connection, dividing the two seas and the southern states of Chiapas, Tabasco, and the Yucatan peninsula from the rest of Mexico. The Uxpanapa resettlement zone occupies the same relative position along the foothills of the upper Gulf slope as Ojitlan, but it has never supported a permanent Indian population and has long been one of the most isolated areas in the region.

History

The next two sections are drawn from the following sources: Brissot, 1837; Carrasco, 1948; Casteneda, 1958; Covarrubias, 1946; Del Corral, 1964; Glick, 1953; von Humboldt, 1817; Iglesias, 1966; McBride, 1923; Ortiz de Ayala, 1966; Ortiz Wadgymar, 1971; Pepper, 1906; Revel-Mouroz, 1972; and Siemens, 1975.

The history of the Isthmus has been closely linked with a bewildering variety of plans and projects to build a permanent link between the oceans. Within a year of the conquest, Cortes sent Gonzalo de Sandoval, conqueror of Tuxtepec, to explore the Coatzacoalcos river. He hoped to find a navigable strait which, as he wrote to his king, "is the thing I wish to meet with more than anything else in the world, for the great service which Your Majesty would derive" (quoted in Covarrubias, 1946, p. 163). A direct water route does not exist, of course, but a port was established at Espiritu Santo across the river from the modern city of Minatitlan. Goods were shipped up the river to Utlatepec, near Serabia, from whence they were taken overland to Tehuantepec. Materials to build the fleet in which Cortes explored California were transported in this way. The route did not last long. Diseases and attacks by English pirates drove out the Spanish, and the crown restricted all trade to the more easily defended ports of Acapulco and Veracruz.

In the eighteenth century, the Tarifa valley, as the Uxpanapa region was then called, was used as a source of timber for the royal shipyards in Cuba. This trade awakened new interest in a transcontinental route. The Isthmus was carefully surveyed in 1774. In 1808, Alexander von Humboldt suggested that a canal, or at least a permanent road, would provide tremendous benefits to the country, but nothing was done (von Humboldt, 1811, vol. 4, pp. 19-27).

After independence, various attempts were made to get a project underway and to encourage the colonization of the region. In 1823 a separate province of Tehuantepec was established. A general issue of free lands in 10 acre plots to war veterans was ordered, but there were few takers (McBride, 1923, p. 90). In 1825 an official named Tadeo Ortiz explored the isthmus. As the rivers on the Pacific slope are not large enough to carry boats of any size, he suggested that a permanent road be built from Tehuantepec through the Tarifa pass to a port on the upper Coatzacoalcos. He was impressed by the potential of the Uxpanapa area (Ortiz de Ayala, 1966, p. 131):

> The interior plains are covered with dense forest and are infested with fevers and diseases, as are all unpopulated regions. Once cleared, however, this area would be suitable for the establishment of colonies. Their produce could easily be transported to market over the several rivers which feed the Coatzacoalcos.

He organized a group of French families in 1829 to open the jungle to foreign settlement. They established coffee plantations on the banks of both the Coatzacoalcos and the Uxpanapa River which were still there 65 years later (Romero, 1898, p. 366), but the experiment did not last long. Their numbers were decimated by accidents on the rivers, diseases, and suicide; the unfortunate Europeans returned home after only three years. Attempts to colonize the upper portions of the basin have continued sporadically for 150 years.

In 1842, President Santa Ana granted a band 300 miles wide from sea to sea to an adventurer named Jose de Garay as a personal concession. He used the generous government subsidies to visit European capitals and promoted the idea of a sea-level canal. During the Mexican war, the United States government attempted to acquire the concession for itself, but eventually dropped the claim. This initiated a series of speculative, diplomatic, and political maneuvers that prevented organized development and put a constant drain on the national treasury for 65 years. Projects ranged from a fantastic canal running through a tunnel to an absurd, triple-tracked railroad on which fully-loaded steamers would be carried on giant, wheeled dry-docks. As ships grew larger, it became clear that it would be impossible to build a canal at reasonable cost.

No fewer than eight different concessions were granted to foreign contractors to build a railroad and port facilities at either end. The job was finally completed in 1907 by Sir Weetman Pearson, a British engineer who had previously overseen the drainage of the valley of Mexico and who founded the first oil company in Mexico. A total of over 40 million dollars was spent on the project, most of which went into the pockets of speculators and corrupt government officials. The line was very successful for several years. The combined sea-land route was cheaper than the transcontinental railroad between San Francisco and New York, and carried large quantities of sugar between Hawaii and the eastern seaboard of the United States. As many as 20 trains a day ran in each direction. The opening of the Panama Canal in 1914 and the disturbances of the revolutionary period caused a precipitous decline in traffic, and the railroad has never made money since. The highway that runs parallel to the line was built in 1946.

Most of the tropical forest in the northern isthmus, including Uxpanapa, was divided up into corporate holdings as railroad grants, 1883 law survey companies (companies deslindadores), and timber concessions in the late nineteenth century. Sir Weetman Pearson, the railroad contractor, and Jose Yves Limantour, President Diaz's Minister of the Treasury, controlled thousands of hectares each. Other important grants were awarded to the Mexico City merchant house of Sanborn's and to various American corporations including William R. Hearst & Company. In theory, these grantees were supposed to survey the land and encourage settlement by small holders. The experience of the American West in the past century might suggest that the construction of the railroad would have led to the clearing and colonization of the surrounding territory. In fact, however, the corporations held onto their land, both to exploit the reserves of mahogany and other precious woods, and on the unfounded expectation that the commercial success of the line would greatly increase the value of their holdings (Revel-Mouroz, 1972).

The expansion of working haciendas was confined to the already-settled zones near the coast of Veracruz and in Oaxaca. Before and during the revolutionary period, there were serious outbreaks in violence between the cattlemen and the Indian communities in the area between Acayucan and Santiago Tuxtla. The local mestizo elite has been powerful enough to maintain much of the good land in private hands down to the present day (Siemens, 1964).

The holdings of the absentee corporations in frontier regions reverted to the federal government during the revolutionary period. Uxpanapa was so isolated from population and commercial centers that not until the late 1960s was the land formally distributed to prevent a new wave of private speculation. Even then, most of the campesinos who were granted rights in the new ejidos were unable to take possession of their isolated grants.

Patterns of Recent Development

Even though large areas have remained almost entirely uninhabited, the northern half of the isthmus has experienced rapid economic growth in the past 40 years. The band of oil fields and petrochemical plants between Minatitlan, Veracruz and Ciudad Pemex, called Chiapas, is the center of the Mexican petroleum industry and produces oil, gas, and a variety of chemical products including fertilizers. Pipelines cross the isthmus and link it with the rest of the country. Including facilities still under construction, the hydroelectric and thermal generating capacity of the region is over two million kilowatts (Mexico, 2). Crude oil reserves were estimated at between 50 and 300 billion barrels in 1978 (Lopez Portillo, 1978). This resource, in addition to sulphur, phosphate, and other mineral deposits, guarantees continued industrial expansion.

This enclave of highly specialized, high-technology industry has not encouraged the balanced development of the isthmus as a whole. Most of the jobs that have been created have been filled by professionals and skilled workers from the other parts of the country. An estimated 87,000 people moved into the region in the 1950s, 75 percent of them to the industrial centers. The average annual population increase of seven percent was one of the highest in Mexico (Mexico, 30). These immigrants receive good wages and enjoy a comfortable standard of living. They consume high-value foods and goods from other regions and from abroad, as Coatzacoalcos is a free port. Local agricultural production has not been stimulated. The illiterate, unskilled campesinos cannot find work in industry, and inflation is constantly pushing up their cost of living.

Over a third of the Coatzacoalcos basin is still virgin rain forest. Productive commercial agriculture is limited to a few modern cattle ranches and specialized enterprises in the Acayucan area. The transition between the most developed and one of the least developed area of the Southeast is abrupt. Hidalgotitlan, the seat of a 1,700 square kilometer municipio, is only 30 kilometers up river from the modern boom town of Minatitlan. It has a population of 2,000, no running water, few other services, and is not accessible by road in the rainy season (Universidad Vercruzana, 1973). The forests to the south, including Uxpanapa, have been coming under increasing pressure. The construction of new roads, the distribution of land in ejidos, the eradication of malaria, and colonization projects of several kinds are extending the area under cultivation. Poor transportation, the lack of well-organized markets, conflicts over tenure rights, and a variety of other problems have prevented orderly development. Subsistence-oriented agriculture based on slash-and-burn techniques and extensive, poorly managed cattle operations are expanding rapidly.

THE RAIN FOREST

The 2.5 billion hectares of rain forest remaining in the world are rapidly being cleared (Dorozynski, 1978). The FAO has estimated that two million hectares are deforested each year in Latin America alone (Nelson, 1973, p. 9). Unlike temperate forests, which are composed of a relatively small number of broadly adapted species, a single hectare of rain forest supports between 50 and 200 different species of trees, plus hundreds of smaller plants, animals, and insects, each one of which is adapted to a very specific ecological niche. Many cannot survive or reproduce themselves once large areas have been cleared and the delicate natural balance has been altered (Gomez-Pompa, 1974). Many species will inevitably become extinct as the pressures of development push back the frontier all over the tropical world. Concern over the future of the global environment has become increasingly focused on this issue in the past decade. (Some recent publications are: Dasman, Milton, and Freeman, 1973; IUCNNR, 1975; and Pearson and Pryor, 1978.) Some scientists have predicted devastating climactic changes, the disruption of the earth's oxygen cycle, and a serious reduction in the gene pool on which future evolution depends if this process is not stopped (Richards, 1973, p. 59).

> . . . the rain forest is retreating. The destruction in modern times of a forest that is millions of years old is a major event in the Earth's history. It is larger in scale than the clearing of the forests of temperate Eurasia and America, and will be accomplished in a much shorter time.

Uxpanapa is part of one of the largest remaining rain forests in Mexico. The resettlement program has excited a public controversy between the Papaloapan Commission and a group of ecologists over how this resource should be managed. The commission, its critics, and the Chinantecs themselves are all very aware that many colonization projects in similar regions have wasted irreplaceable resources without achieving their economic or social goals. The government is committed to the policies of increasing agricultural production in rain-fed areas, creating sources of rural employment to stem the tide of migration to the cities, and promote regional development. The Uxpanapa project is an attempt to use the resources of the tropical environment to meet these goals. It will have lasting effects, not only on the Chinantec people, but also on the future development of the forest frontier in the Southeast.

Geology and Climate

The Federal Drainage District of Uxpanapa is located just north of the 17th Parallel in a depression that runs from west to east along the foothills of the Chimalapa mountains. Separated from the plains to the north by a low range of hills, it is 75 kilometers long by 35 wide, and covers an area of 260,000 hectares. The bed rock consists of marine sediments, and is predominantly dolomitic limestone. This material weathers rapidly and is highly permeable. Apart from the six rivers that cross the zone on their way from the mountains to the Coatzacoalcos and a few brooks, surface water is limited to natural cisterns, called cenotes. Recent deposits of gravel and silt have built up along the margins of the rivers.

The topography is irregular, ranging from a nearly level central strip to steep hills and frequent limestone outcroppings. Altitude varies between 100 and 150 meters above sea level. Sixty percent of the land is on a slope greater than 15 percent, which is too steep for agricultural use without serious danger of erosion.

Precipitation varies considerably over short distances, depending on the elevation and the orientation of the hills to the prevailing winds, and is greatest at the eastern end. Annual rainfall varies between 3,000 and 4,000 millimeters (120-160 inches). There is a four-month relatively dry season between February and May when 10-15 percent of the rain falls (figure 6.1).

Average monthly temperatures range between 21 and 29 degrees centigrade, with highs of 40° in the dry season and lows of 10° in the winter. Relative humidity is very high during the entire year, averaging 70 percent in May and over 90 percent in November. Over two-thirds of the days are cloudy. The Isthmus of Tehuantepec is subject to complex wind patterns. The prevailing northeasterly trades bring moist air in from the gulf during the rainy season. Strong, cool winds from the north, called nortes, blow in the winter months.

Weather stations have been established too recently to provide detailed information on the long-term climatic patterns. Data from one of the new stations are summarized in table 6.1. According to the Koeppen system as modified by Garcia for Mexico, most of the region can be classifed as Af (i') g. (This is a hot, humid climate, where the mean temperature in the coldest month is above 18C., more than 10

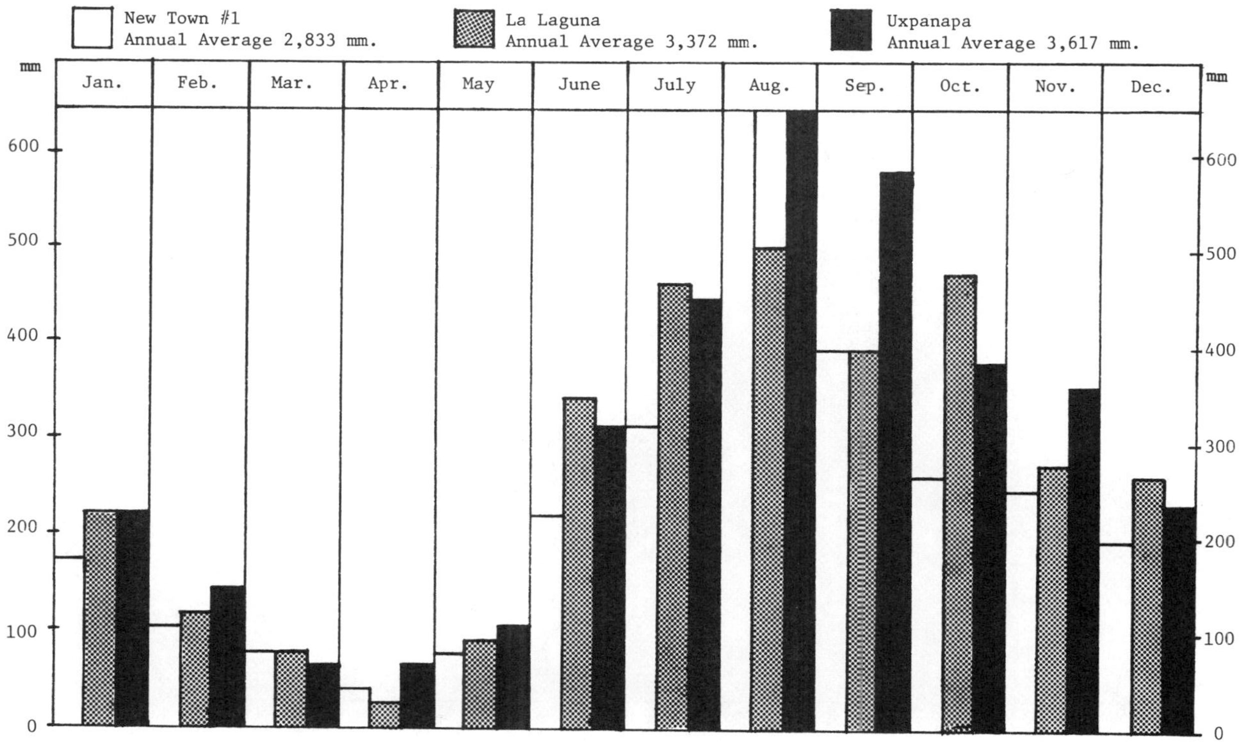

Fig. 6.1. Uxpanapa: monthly rainfall pattern - three-year averages from selected stations, 1974-1977.

Source: La Comision del Papaloapan.

Table 6.1. Uxpanapa: Data from Weather Station at La Laguna

Year / Month	Temperature: Maximum	Temperature: Minimum	Temperature: Mean	Precipitation: Mean, Daily	Precipitation: Total	Total Evaporation	Number of Days: Clear	Number of Days: Rainy	Number of Days: Cloudy
	(degrees centigrade)			(millimeters)		(millimeters)			
1975									
July	34	21	26.7	11.9	370.0	104.90	7	21	24
August	33	20	26.3	22.5	699.0	150.08	6	24	25
September	34	19	25.2	13.9	413.6	104.25	1	26	29
October	35	15	24.1	20.8	645.2	97.80	2	21	29
November	32	14	22.7	8.1	241.8	68.38	9	20	21
December	32	4	21.2	6.0	186.3	78.53	5	19	26
1976									
January	33	11	19.2	12.6	391.0	55.40	6	21	25
February	34	11	20.3	3.5	101.9	82.34	9	11	20
March	37	14	25.7	2.5	77.5	125.77	18	10	13
April	36	15	26.4	.4	118.0	153.40	22	8	8
May	38	20	27.7	3.7	115.1	156.75	19	8	12
June	38	20	27.4	6.9	208.1	119.49	18	14	12
TOTAL					3,567.5	1,297.09	122	193	244

Source: La Comision del Papaloapan.

percent of the annual rain falls in the winter, more than 60 mm. of rain falls in the driest month, variation between mean monthly temperatures is between five and seven, and the hottest month of the year comes before June. Portions can be classified as AM(f)(i') g, which means that there is less rain in the rainy season. According to the Thornwaite system, most of the area can be classified as ArA' a.)

Vegetation

Uxpanapa supports a tremendous variety of plants. In a preliminary study, nearly 500 species were counted in a randomly chosen circle with a radius of 15 kilometers (Caballero Nieto, 1976). The variations in climate and soil type are reflected in the composition of the forest. Portions of the area have been cleared over the years, some of them repeatedly, so the existing jungle is a mixture of climax and secondary associations. The vegetation has been classified into five broad groups (Echenique-Manrique, 1975; Gomez-Pompa, 1974).

Tall, Evergreen Forest

The most common association is the classic, three-tiered, hardwood forest found on deep, well-drained soils, both in flat areas and on hills. The highest trees are 25-40 meters tall; the middle level, 12-25; and the lowest, 5-10. Because the nutrients are concentrated in the top layer of the soil, even very large trees put out most of their roots horizontally, and many have developed braces and buttresses to support themselves. A wide variety of vines, epiphytes, and other small plants find niches at various levels within the system where they can compete for air, light, and water.

Tall, Semidecidous Forest

This association shares many species with the first and is similar in its structure, but 20 to 50 percent of the trees shed their leaves during the dry season to prevent excessive transpiration and dehydration. It is found on the limestone outcroppings and in other areas with shallower, less fertile soils with the lower water-holding capacity, or where the local climate is drier.

Riparian Forest

Along the banks of the rivers on the better alluvial soil, which are periodically disturbed by flooding, the forest is more diverse and is not divided into distinct levels. These areas are suitable for crop production during the dry season.

Bamboo Groves

This is a very interesting association from an ecological point of view. It is found in areas cleared in the last century during the foreign lumbering operations, or perhaps even earlier. Bamboo is an aggressive plant with a very dense root system which has maintained itself in almost pure stands on soils with a high clay content, very little organic matter, and low fertility.

Secondary Associations

When the forest is cleared and then left to fallow, fast-growing and aggressive plants, which are collectively called acahual in Mexico, quickly invade. Within a year, the land is completely covered with a dense stand of light-loving bushes, grasses, vines, and legumes two or three meters tall. These plants are much better adapted than any economic crop to depleted soils and high rainfall. The development of the secondary sequence is very complex. Over 100 years would be required to reach the climax, and many species will inevitably become extinct as the forest recedes.

Fauna

Uxpanapa, like all tropical forests, supports a limited variety of species of large mammals and birds. As the area is settled, the tigers, tapirs, deer, armadillos, monkeys, and other animals are being hunted down by the campesinos and workers. The bird population is increasing in both number and diversity with the changing ecological balance and the introduction of crops. The effects of this process on the insect and microorganism population have not been studied.

Soils

A comprehensive soil survey has not yet been made. The basic planning decisions that determined which areas would be cleared by machine, where the roads would be built, and where the new towns would be located were made on the basis of a preliminary survey made in 1975 from aerial photographs taken before the forest was cleared (Mexico, 21). The zone was divided into seven broad categories on the basis of topography, which are summarized in figure 6.2. Ground checks and chemical tests were few, large areas of stony ground did not show up, and the area suitable for mechanized agriculture was overestimated by over 100 percent. In 1977, after 10,000 hectares had been cleared by machine, a more detailed study was made of a much smaller area (Mexico, 11). The results are also summarized in figure 6.2.

Both reports emphasized the preliminary nature of their results and concluded that Uxpanapa was more suitable for perennial tree crops and

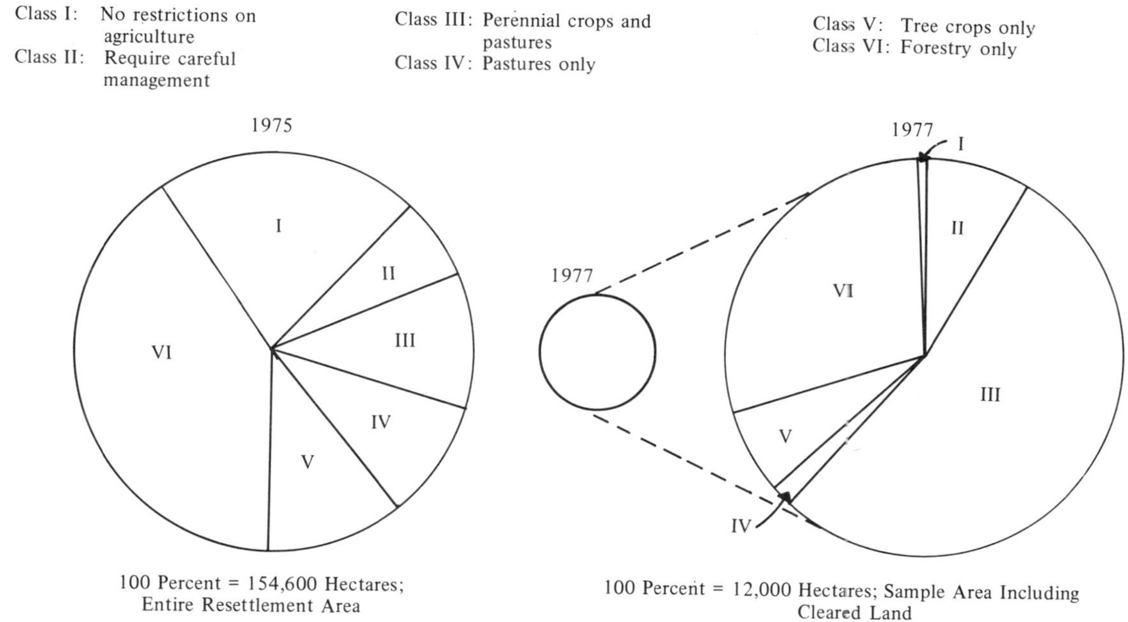

Fig. 6.2. Uxpanapa: Two approximations of the productive potential of the land.

Sources: Mexico, Secretaria de Agricultura y Recursos Hidraulicos, Comision del Papaloapan, Agrologia Desarrollo, S.C., "Estudio Agrologico Semi-detallado - Distrito de Drenaje de Expanapa," mimeo, 1977; Mexico, Secretaria de Recursos Hidraulicos, Comision del Papaloapan, Estudios y Proyectos, S.A., "Reconocimiento Agrologico Para el Desarrollo Agropecuario de Uspanapa," mimeo, 1975.

pastures than for annual grains.

Over the centuries, the consistently hot, rainy conditions have produced highly weathered soils with an average depth of between two or three meters. Samples have been classified as ferrasols and rhodic ferrasols according to the FAO system. These correspond roughly to oxisols and alfisols in the USDA system. Although the parent material is primarily limestone, which provides calcium and other bases, high leaching rates have produced acid soils with pH values between four and six. Higher values have been found in areas immediately surrounding the limestone outcroppings. The soils are 40 to 60 percent kaolinite and hallosyte clays, which lock together in a stable, permeable structure through which water drains very easily. Iron and aluminum, which weather much more slowly than silica and other materials, have accumulated to produce a characteristic yellowish-red color. Most of the available water and plant nutrients are stored in the top 30 centimeters, where organic matter has accumulated. A series of tests were made in 1977 in two pilot areas about 15 kilometers apart, the results of which are summarized in table 6.2.

Broadly similar types of these ferrasols cover large areas of the zone, but their specific properties vary considerably depending on the parent materials, topography, drainage patterns, and other factors. The soils on even gentle slopes of 5-7 percent are noticeably thinner, less fertile, and more prone to erosion than the flat lands. Some of the alluvial soils near the rivers are gravelly and poor, but a few highly productive plots have been identified and cultivated by slash-and-burn cultivators over the past ten years. Field drainage will be necessary only in a few small areas.

The soils' most obvious advantage in a high-rainfall area is their strong, high permeable structure. Heavy machinery can move over the fields within hours of a torrential downpour without getting bogged down in mud. This rapid and efficient drainage is a mixed blessing. As little as a week without rain during the growing season can cause a mini-drought and put the crops under stress. Applied fertilizer, especially nitrogen, leaches out very quickly and must be reapplied. The water-holding capacity will decrease even further should the organic-matter content be allowed to decline.

The high clay content means that compaction can be a problem, even though the soils are very strong and a laterite layer cannot form. The surface is very delicate when it is first uncovered. Splash, sheet, and gully erosion can develop very quickly if the land is not immediately replanted.

The complex changes in soil fertility and structure, which inevitably accompany the clearing of a forested area, have not been monitored. Once the land is exposed to the sun, higher soil temperatures increase the rate of microbial decomposition of the organic matter into soluble forms. If it is not replaced, the high rainfall leaches the nitrogen and other nutrients down through the porous soil out of reach of the roots of the crops. The crystalline structure of the clays fixes phosphorus very tightly, so that it is not available to the plants. The status of the

Table 6.2. Uxpanapa: Chemical Analysis of Soil Samples, 1977

Test	Characteristic	Percent of Samples Area 2	Percent of Samples Area 3
pH	Strongly acid (4-5)	23.2	7.4
	Moderately acid (5-6)	60.7	74.0
	Slightly acid (6-7)	16.7	16.0
O.M.	Very poor (less than 1%)	28.6	15.0
	Poor (1-2%)	67.9	61.0
	Medium (2-3%)	3.6	25.0
Total N	Very poor (less than .05%)	28.6	14.8
	Poor (.05-.10%)	67.8	61.1
	Medium (.10-.15%)	3.5	25.9
Available P	Poor (less than 11 ppm)	---all samples--	
Available K	Rich (146-222 ppm)	16.0	85.0
	Very rich (more than 222 ppm)	84.0	15.0
Available Ca	Medium (490-1110 ppm)	78.0	85.0
	Rich (1110-1780 ppm)	20.0	9.0
	Very rich (more than 1780 ppm)	2.0	6.0
Available Mg	Very rich (more than 90 ppm)	---all samples--	

Methodology: 182 samples of the top 30 centimeters of soil

pH - H_2O - water:soil; 2:1
O.M. - (%)Walbley and Black
Total N (%) Kjeldahl
Available P (ppm) Troug
Available K (ppm) Dilution with ammonium acetate, flame photometer
Available Ca (ppm) Titration with triethenolamine
Available Mg (ppm) Titration with triethenolamine

Source: Mexico, Secretaria de Agricultura y Recursos Hidraulicos, Comision del Papaloapan, Agrologia Desarrollo, S.C., "Estudio Agrologico Semi-detallado - Distrito de Drenaje de Uxpanapa," mimeo, 1977.

micronutrients that are necessary in small amounts require investigation. Isolated samples taken under different conditions and analyzed by different laboratories can produce very misleading results. So few surveys have been made in tropical Mexico that there is no reliable basis for comparison. There is an urgent need to carefully analyze and map the soil resources of Uxpanapa.

SPONTANEOUS COLONIZATION

Uxpanapa has not supported an indigenous population for many hundreds of years. Pieces of pottery and other artifacts were turned up in the clearing process, but there is no evidence that the zone was ever more than a sparsely settled frontier. Nevertheless, there were about 200 families living in some 40 scattered settlements when the commission arrived. Almost all of them had moved into the region in the previous five or ten years as part of a very loosely organized government colonization program. Poor and uneducated campesinos from various parts of Veracruz and Oaxaca, they were farming with the hand tools and seeds they had brought with them.

These people are an example of how an isolated tropical area is spontaneously colonized in the absence of the kinds of infrastructure and services the resettlement program is providing to the Chinantecs. They did not have any draft animals, much less chain saws, tractors, or other mechanized implements. Communication with the outside world was by river and forest trail. They cultivated very small plots, and were not able to meet their needs from agriculture alone. Their incomes and the returns to their labor were extremely small by any standard. Nevertheless, on the basis of only a few years of experience, they were able to produce maize, rice, and other crops. They chose their plots, varieties, and practices with care. Although the yields were low, they were higher than the average that the expensive, mechanized systems introduced by the government were able to achieve in the first years of the new project. Their experience provides a valuable perspective on the challenge of introducing extensive agriculture technology into the rainy tropics.

The Formation of the Ejidos

From the air, areas of secondary vegetation along the margins of the rivers stand out clearly from the surrounding forest. This implies that Uxpanapa had been intermittently inhabited over a long period of time. Since the end of the timber boom in the early twentieth century, it has been one of the most isolated regions in the Southeast. A few airstrips were built to bring in wealthy hunters and to ship out marijuana, but permanent settlements were not formed until the mid-1960s. At that time, the Agrarian Affairs Department (DAAC) divided the zone into over 100 ejidos and granted rights to several thousand names from their lists of petitioners for land (Mexico, 4). The government was anxious to create ejidos at least on paper as quickly as possible to prevent unauthorized settlement and land speculation. (Only one small area, in the valley of the Uxpanapa river, was privately owned. It belonged to a consortium of ranchers from Chiapas, who hired peons to clear and prepare the land for the introduction of cattle. It was expropriated for divison into ejidos in 1976.) Field surveys were minimal and little more than the cost of transportation was provided to the colonists.

The agrarian issues of land tenure and ejidal rights are always complicated in Mexico. Many petitioners were not legitimate farmers and had no intention of moving into the wilderness. Others were granted parcels in totally inaccessible areas. According to the law, each ejidatario must clear some of his land and occupy it within two years to validate his claim. Migrant cane cutters, petroleum workers, and others without legal rights were either hired by the grantees to establish residency, or came on their own account. The resulting legal confusion complicated the development of orderly communities. The settlers who had the greatest success moved in together as an organized group under the direction of a recognized leader. By 1974 only those ejidos in the most accessible and fertile areas were occupied. The Chinantecs are being moved into the vacancies in this structure.

Patterns of Settlement

The colonists looked for flat areas near the river with access to fresh water for their settlements. Most of them embarked on the Solosuchil river at Jesus Carranza, an old railroad town and regional center. The water level in this and the other two rivers providing access to the zone fluctuates widely depending on the rainfall in the mountains. A trip that took seven or eight hours between July and January could require several days in the dry season, when the boats had to be dragged over innumerable shallows. In some years, the villages were completely cut off from the outside world for two or three months. Private merchants moved up and down in launches at irregular intervals to trade tools, clothing, and other basic goods for agricultural produce. They paid 20 to 50 centavos per kilo of maize, less than a quarter of the support price paid by CONASUPO, the government buying agency. Fumigation teams from the malaria eradiction campaign came through from time to time, but there was no other health care, no schools, and no agricultural credit or extension.

The small communities of two or three dozen families were entirely dependent on what they could extract from their environment for their survival. They hunted for meat, and for tiger skins and exotic feathers they could sell. They fished and gathered barbasco. They cleared small areas of forest to meet their immediate needs and to begin the development of permanent settlements.

Slash-and-Burn Systems

The most productive plots were located in low-lying areas along the rivers, which were convenient to the settlements and relatively easy to clear because periodic flooding kept the trees from growing tall. Except in gravelly areas, the alluvial soils, formed from sediment brought down by the rivers, are fertile, although they became seriously water-logged during the rainy season. During the dry winter cycle, these areas produced the highest yields in Uxpanapa of maize, beans, and chili.

After the trees were cut down, the following system was established: Leguminous weeds and grasses were allowed to grow through the wet summer months and were cut in late October and November. The trash was not burned, but was spread evenly over the plot as a layer of mulch that both fertilized the soil and protected it from drying out. Maize was planted directly through this mat in December with a planting stick. The heavy rains last into January, which allowed the plants to get a good start. During the rest of the growing season, the crop depended on the residual moisture held in the soil pores and conserved by the mulch, supplemented by an occasional shower. Between 1.5 and 3.5 tons per hectare were harvested from small plots using this system.

These alluvial areas are small and can only produce one crop a year. The ejidatarios also cleared the high forest that grows on deep ferrasols, where much more work produces less return. It is really too wet to grow maize profitably in Uxpanapa during the rainy season. The plant suffers from excess moisture and the lack of direct sun, and is attacked by a great variety of pests and diseases. The soil structure promoted very efficient drainage, so that a week without rain puts the crop under stress. The farmers spread their plots out over different kinds of soil and planted different native varieties at different times to minimize their risks. Yields averaged between 900 and 1500 kilos per hectare.

Upland rice was the single most important crop grown primarily for sale. The alluvial soils are too wet to work in the rainy season, so the crop was put in the first year after clearing a plot of high forest. Residual soil fertility was high. Weed, insect, and disease damage generally fell within acceptable levels until the second or third season. Relatively tall native varieties such as Lira and Morada Criolla were vigorous enough to out-compete weeds in the early stages of growth. Yields varied between 600 and 1800 kilos per hectare, which was not enough to justify the use of fertilizers or chemical pest control.

Chili peppers, beans, sesame, and a few other crops were also grown for sale in small quantities, but both yields and markets were very uncertain. Cattle, bananas, pineapples, fruit trees, and other crops are being introduced as the area is opened up by new roads.

The Returns to Labor

By almost any measure, the colonists made very little from the extremely hard work required to clear virgin forest and grow enough food to survive. The middle range of labor requirements and yields for the principal crops were as follows:

Crop	Ferrasols	Alluvial Soils
Summer Maize (Temporal)		
Yields	150-1500 kgs/ha	
Man-days/ha	50-60	
Winter Maize (Tonalmil)		
Yields		900-3500 kgs/ha
Man-days/ha		44-55
Rice		
Yields	600-1800 kgs/ha	
Man-days/ha	65-75	
Beans		
Yields		200-1400 kgs/ha
Man-days/ha		65-75
Chili		
Yields		1000-5000 kgs/ha
Man-days/ha		90-125

According to purely economic criteria, both the average and the marginal returns to labor were extremely low. Even in Mexico, where it is estimated that half of the population is either unemployed or underemployed, the settlers probably could have earned more in the first years by migrating to the city rather than into the forest. Energy flow studies, which balance the number of kilocalories expended in agricultural labor against the food energy value of the product, have been suggested as a better way to measure the efficiency of labor in societies that operate outside of a market structure. An elaborate study of a Neolithic people in an isolated valley in New Guinea traced complex interrelationships between the fertility of the soil, the requirements of a slash-and-burn system, and the human and livestock populations (Rappaport, 1971).

A biologist used this method to measure the ecological efficiency of the cropping systems used in an ejido in Uxpanapa (Caballero Nieto, 1976). He observed the tasks involved in cultivating the principal crops, and estimated the human energy requirements. Using laboratory techniques, he measured the energy content of a kilogram of each crop, and multiplied this figure by the average yield. He was then able to derive energy input-output ratios, which he correlated with three major ecological subzones. The results are summarized in table 6.3.

Soils that supported small trees such as bamboo were relatively easy to clear, but were highly acidic and produced low yields. The better soils were heavily forested, and required so much energy to clear that the energy ratios were not significantly better. The ejidatarios were not entirely isolated from the market and did sell some of their harvest. A reasonable economic return depended on the intensive production of relatively high value crops such as chili and sesame, which faced fluctuating prices and uncertain markets. The biologist concluded that traditional crops and agricultural methods were neither an efficient use of labor nor an appropriate use of the resources of the tropical forest.

There are several conceptual problems with this approach, even assuming that the data are reliable. Even in an isolated, primitive society, the efficiency of human effort is not an easy parameter to tie down. The issues are much more complex in a rapidly changing pioneer settlement. Kilocalories measure values not evident to a Mexican campesino. Sesame, an oil seed, has a very high energy content, even though its market price is low. Chili peppers are mostly water, but they bring a high economic return if a market can be found. It is difficult to argue that a person chooses crops, soils, and cultural systems to produce the most food energy with the least effort.

The empirical knowledge on which slash-and-burn systems are based has been accumulated over thousands of years. The spontaneous settlers in Uxpanapa drew upon this tradition to support themselves in the wilderness, but they did not expect the hardships of the first few years to last forever. They had no intention of establishing stable agricultural systems in isolation from the rest of the country. Once they had justified their claim to the land by clearing it, they had every reason to believe that a road would be built and that they would be able to evolve more profitable enterprises. If the area had not been chosen to resettle the Chinantecs, it probably would have developed along the same lines as other regions of the Southeast. Experience would suggest that many of the original colonists would have been pushed out by cattle ranchers as communications improved and the fertility of the land declined.

The spontaneously settled ejidos have benefited significantly from the activities of the Papaloapan Commission, particularly from the roads and the availability of wage labor. Nevertheless, there have been many conflicts with the complex bureaucratic structure. Many of the colonists believe that the system of credit and mechanized agriculture being introduced provides them with a lower return than they could achieve by improving their traditional methods. They also feel that their hard-earned independence is threatened by a program set up to meet the needs of the Chinantecs and of the national economy, rather than their own.

Table 6.3. Ejido "Augustin Melgar," Uxpanapa: Energy Balance Sheet

Ecological Zone (Predominant Soil Type)	(A) Crop	(B) Labor per Hectare	(C) Kilocalories Expended per Hectare	(D) Average Yield	(E) Kilocalories in Crop per Hectare	E/C
		(man-days)	(thousands)	(kilos per hectare)	(thousands)	
High Jungle (Ferrsaols)	Summer					
	Maize	54	203	1,711	6,269	31
	Rice	70	247	1,633	5,792	23
	Sesame	51	204	1,066	6,354	31
Low Jungle (Thin Ferrasols)	Winter					
	Maize	54	192	1,632	5,975	31
	Beans	73	263	1,440	4,608	18
	Chili	99	283	2,100	589	2
Bamboo Groves (Sandy Loams)	Summer					
	Maize	38	139	382	1,400	10
	Rice	58	192	1,053	3,735	20
	Sesame	39	144	1,066	6,354	44
	Chili	129	265	2,175	579	2

Source: Javier Caballero Nieto, "El Costo Ecologico del Uso de la Tierra en un Ejido del Tropico Mexicano" (Mexico: Tesis Profesional, Departamento de Biologia, UNAM, 1976).

7 "We Will Be Realists and Do the Impossible"

This motto is inscribed on the arch the Papaloapan Commission has built at the entrance to Uxpanapa. The explicit purpose of the project is to conquer a hostile environment to meet national goals. Over the past forty years, the Mexican government has undertaken many agricultural development projects. Almost all of them have involved the irrigation of arid lands in the north. They have been administered by engineers, and have emphasized the construction of dams, canals, roads, and other infrastructure. The country has become a leader in this field, and exports the technology.

The same basic priorities were applied in Uxpanapa. By concentrating capital, machinery, population, and services, the managers hoped to simplify the natural and social factors involved in tropical development so that they could be effectively coordinated within a single, centralized framework. This emphasis was supported by powerful groups, including construction companies and other contractors. It fit the criteria that until recently dominated the lending policies of the international agencies, although this project has been funded entirely from Mexican sources.

Although the technology involved in bringing a large area in the rainy tropics into intensive production had not been tested, the Papaloapan Commission was under heavy pressure to organize the project as rapidly as possible. The Cerro de Oro dam was due to be completed in 1978, and the Chinantecs had been promised that a complete infrastructure would be ready for them to move into before their lands were flooded. At the same time, the national production crisis required all agencies of the federal government involved in agriculture to increase the production of basic food grains on a crash basis. Preliminary studies of the natural environment were made, but there was not time to organize a careful experimental program before construction and clearing began.

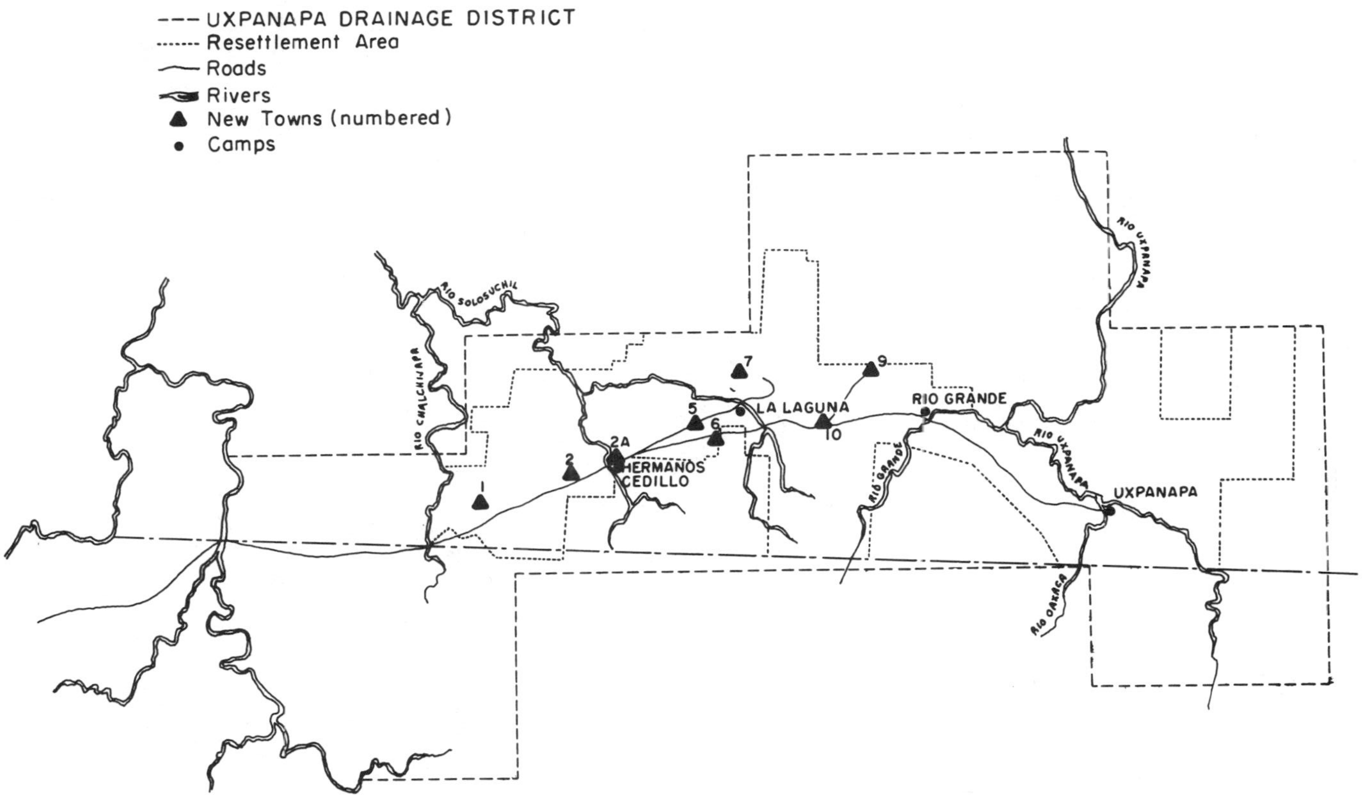

Fig. 7.1. Uxpanapa resettlement zone.

THE BASIC PLAN

Work got underway as soon as President Echeverria approved the proposals the commission presented to him in 1974, and which can be summarized as follows:

1. Each resettled family will receive rights to 20 hectares, but the land will not be individually assigned. The people will be organized into collective ejidos, each one of which will plant a combination of annual and perennial crops in compact blocks.

2. A main road running the length of the zone will be fed by a regular grid of secondary roads, providing rapid communications between the fields, the population centers, and the markets for inputs and products. The road will eventually be connected with the interstate systems, at which time Uxpanapa will be well located as a regional center.

3. As much marketable timber as possible will be extracted before the land is cleared for agriculture.

4. The land suitable for mechanized agriculture will be cleared rapidly by machine, permitting large areas to be brought under cultivation immediately. Areas with a slope of greater than ten percent, or with other limitations, will be cleared by hand for pastures, rubber plantations, and other perennial crops.

5. The population will be concentrated in 12 new towns located at regular intervals on soils not suitable for agriculture. The people will build their own houses with a 20,000-peso subsidy from the government. Services such as electricity, running water, sewage, schools, and health clinics will be provided.

6. Tractors and other large, indivisible capital units can be used efficiently on the large, contiguous fields. Mechanization will allow the campesinos to farm a much larger area than they possibly could using traditional methods. Rapid land preparation will allow two crops a year to be grown, reducing seasonal unemployment. As each family will not be dependent on one small area, crop rotations and other conservation methods can be introduced.

7. Central planning under the authority of the Papaloapan Commission will allow the efficient allocation of machinery, seed, fertilizer, and other inputs among the ejidos. Technical assistance can be provided much more effectively to organized groups than to individuals working alone. The choice of crops can be coordinated with national needs, available markets, and the results of experimentation.

8. Credit will be administered by a central office of the official Bank. Each ejido will receive credit as a unit through its elected president, who will be responsible for the distribution of the work and the income among the members of the community. The procedures of

the Crop Insurance Company will be simplified in these large units.

9. A Union of Ejidos will combine resources from all of the ejidos, including the indemnification from the expropriation of the land and common property in the Cerro de Oro lakebed and the income from the timber operation. It will use this equity to obtain credit to invest the profit-making enterprises on a scale that will benefit the zone as a whole. These will include a central machinery station, trucks, buses, a gasoline station, and so on. As the region develops, the Union will invest in secondary industries linked to the agricultural production of the ejidos. These may include a rice mill, lumber mills, and rubber processing plants. As much of the economic activity in Uxpanapa as possible will be owned by the ejidatarios through this structure, limiting the appropriation of surplus value by outsiders.

10. Each family will be given a 1,600 square meter plot of land behind their house. The Commission will encourage women and children to participate in a program of intensive food production for local consumption in these gardens. Central breeding stations will provide pigs, chickens, and rabbits to be fattened at home. Wastes from the gardens and the commercial fields will be processed into feed. A direct marketing system will be set up to avoid the establishment of intermediaries.

The basic infrastructure and the relocation of the Chinantecs will cost approximately 120 million dollars by the time the project is completed. A preliminary breakdown is outlined in figure 7.2. By simple calculation, this amounts to $40,000 per family. It is important to note, however, that the 3,000 families for which the commission is directly responsible will require only 60,000 of the 260,000 hectares in the zone. Once these people are fully installed and their legal position is secure, the remaining area will be opened up to colonization by campesinos from other parts of Veracruz, at a much lower per-unit cost. Indemnification for the land and property flooded by the Cerro de Oro peoject will be charged as a cost of the dam. The infrastructure in Uxpanapa is considered as a long-term investment in the productive capacity of the tropics, and its cost will not be recovered from the project itself. Large, centrally organized, and well-funded, Uxpanapa has avoided many of the problems of the Temazcal resettlement program. The direct benefits to the Chinantecs have been substantial. They have been paid generous indemnification. Their communities are being moved intact to a new area with a much higher level of services than they enjoyed in their homeland. Most of the mestizo class that exploited them has been left behind. Credit, technical assistance, and machinery have been provided. In the 1950s, the government met its legal obligation to the Mazatecs by giving them land, a few basic but generally inadequate services, and leaving them to subsist using their traditional technology. The agricultural programs were experimental at best. In the 1970s the priorities have been reversed. The resettlement program is being used as an opportunity to develop modern agricultural

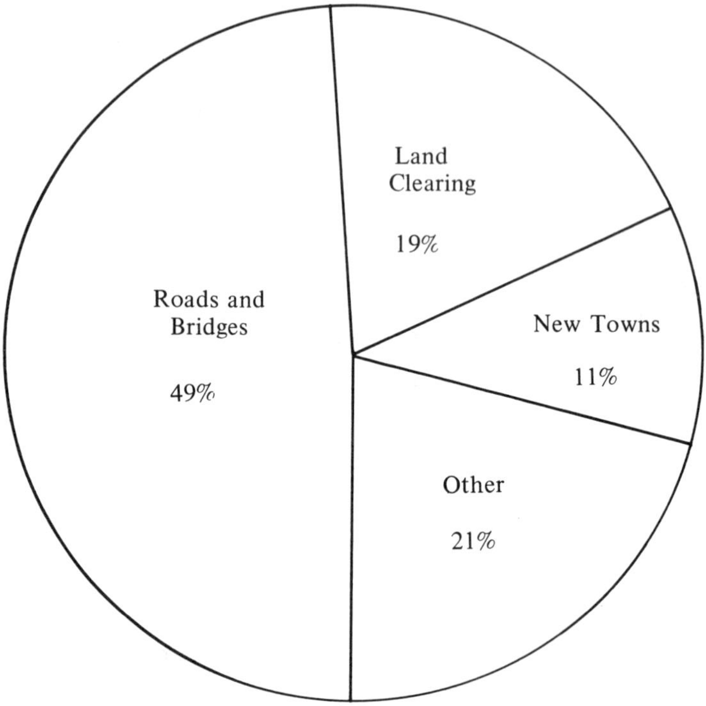

Fig. 7.2. Uxpanapa: distribution of capital expenditures through 1976.

Note: This distribution is based on a total expenditure of 630 million pesos through November, 1976. For obvious reasons, road construction was by far the largest single cost in the first stage. When the project was first planned in 1974, it was estimated that the total cost of the infrastructure would be 1.5 billion pesos, or $120 million at the old exchange rate of 12.50 pesos per dollar. Both the cost of clearing and the projected cost of the Cerro de Oro dam nearly doubled immediately after the devaluation in September, 1976. As of 1978, the Commission had not publicly released a revised, comprehensive budget.

Source: Mexico, Secretaria de Recursos Hidraulicos, Comision del Papaloapan, "Memoria de al Construccion de la Obra en Uxpanapa" (mimeo, 1976).

enterprises under the direct control of the state, organized through collective ejidos.

THE ECOLOGICAL CONTROVERSY

A group of biologists from the National University of Mexico and the Field Museum of Chicago, under the direction of Dr. Arturo Gomez-Pompa, made a study of the Uxpanapa forest in 1974. They have been very critical of the project's emphasis on rapid development and of what they see as the poor quality of its scientific basis. They form part of an influential group of scientists who are trying to reorient tropical agricultural development policy away from large, capital-intensive projects in favor of flexible, small-scale, locally controlled systems based on ecological principles. (The most important organizations dedicated to this goal are the Biology Institute of the National University, INIREB in Xalapa, Veracruz, and the Eco-Development Center in Mexico City. The Uxpanapa study, among others, was funded by CONACYT, Mexico's equivalent of the National Academy of Sciences.)

As botanists, they were very concerned that the diversity and the equilibrium of the tropical environment not be destroyed. They were worried about erosion, downstream silting of the rivers, and the more subtle effects on the region as a whole if the natural systems were thrown out of balance. They saw many examples from all over the tropics of lands that had been opened up to agriculture and ranching, using technology imported from temperate regions. The introduction of monocultures resulted in the deterioration of the soils. After a few years, the areas were only capable of supporting very small populations at the subsistence level, and did not produce surplus commodities in any quantity (Echenique-Manrique, 1975, p. 11):

> It is obvious that the country must produce more food, and that the tropics should provide part of it. Nevertheless, we feel that it is neither sensible nor practical to attempt to "dominate" or "conquer" nature. Man is a living being who is himself part of the ecosystem, and his actions should be directed toward the maintenance of the equilibrium at the same time as he derives resources from it.

Both the Papaloapan Commission's engineers and the ecologists were very excited by the opportunity Uxpanapa presented. The resettlement program provided both a reason and a budget to move into an empty area and develop it according to a single, integrated plant. Both groups had been frustrated in other projects by the complexities of local power structures, land-tenancy arrangements, corruption, and limited funds. The construction schedule of the dam meant that something had to be done quickly, before experimentation and experience provided clear guidelines for action. Both were strongly committed to finding a way to improve the living conditions of the Chinantecs and to ensure their participation in a program of economic development. Both saw the project as an opportunity to develop a model that would allow the productive use of large areas in the Southeast. Nevertheless, their priorities were diametrically opposed.

The Ecologists' Proposals

This section is based on the following sources: Caballero, 1976; Echenique-Manrique, 1975; Gomez-Pompa, 1971; Gomez-Pompa et al., 1974; Mexico, 8; and Toledo et al.

The ecologists looked to traditional agricultural systems as a place from which to start. On the basis of a study in an ejido in Uxpanapa, they constructed table 7.1 as a rough indication of the untapped potential of the tropical environment. Traditional agriculture is almost by definition well-adapted to the local ecological conditions, and takes advantage of a wide variety of resources. Nevertheless, the campesinos produce significant marketable surpluses of only a handful of crops. The biologists identified hundreds of species with potential value as food, fiber, oil, wood, and other useful products in Uxpanapa. They suggested that the forest be cleared in a patchwork pattern, so that each resettled family could grow enough to eat while an intensive program of research and development into new products was undertaken.

They gave several examples of how campesinos and scientists could work together to create economically variable technology. The ojite tree (Brosimum alicastrum sw.) is common in the zone and has been used by the Mayas for centuries for a variety of purposes. The wood is of very high quality and is similar in its properties to maple, which Mexico imports in large quantities. Its prolific fruit is 13 to 15 percent protein and is high in vitamin C. The leaves are also high in protein and would make an excellent cattle feed component if economical harvesting, processing, and feeding methods could be developed. The sap and seeds have long been used as herbal medicines.

Bamboo is one of the few plants found in dense, pure stands in the area. The National University investigated its possible use as a source of cellulose for paper, which is also imported into Mexico. They developed a prototype for a clean, nonpolluting, thermo-mechanical process that used the lignin from the stalks themselves as fuel. Each small plant would require an initial investment of three million pesos, would employ 35 permanent workers, and could produce about 3,000 tons of cellulose a year. Bamboo grows very quickly, and methods could almost certainly be found to propagate it artifically.

The biologists were confident that with enough time, energy and financial support, a great many variable enterprises based on the natural resources of Uxpanapa could be developed. Crops and techniques from other parts of the tropics could be introduced after

Table 7.1. Number of Potentially Useful Species in a Tropical Ejido

	Natural Environment			Transformed Environment			
	Primary Forest	Secondary Forest	River	Village	Garden Plots	Fields	Total
Plants	100	80	1	-	78	11	270
Birds	25	18	9	-	13	13	50
Mammals	20	18	0	-	8	0	29
Insects	4	1	3	-	0	1	9
Fish	-	-	13	-	-	-	13
Reptiles	0	0	7	-	0	0	7
TOTAL	149	117	33		99	25	378

Source: Victor Manuel Toledo, "Uso Multiple del Ecosistema, Conocimiento Tradicional, y Estrategias de Eco-Desarrollo" (Mexico: Instituto de Biologia, n.d.).

careful testing. The natural ecosystem would be adjusted and altered to meet human needs, but it would not be irreparably destroyed.

Each ejido could develop a variety of enterprises including crops, livestock, and one or more specialized small industries at a rate determined by the interests and capabilities of its members, and by the technology that became available. As the region as a whole developed, the ejidos could move from a subsistence to a commercial orientation and could cooperate in larger industries. In addition to the immediate benefits of this process to the Chinantecs, its scientific value would be great. Uxpanapa could become a laboratory for the development of appropriate technology for the rainy tropics.

The biologists felt that the Papaloapan Commission's program was short-sighted. In their opinion, the technical studies were superficial and intended as the basis for a rapid and radical transformation. The possibilities of sustained-yield management of the timber and other forest resources were not considered. The area would be cleared before anyone could assess the danger of adverse, long-term effects.

Valuable timber would be sold at a fraction of its potential value before its best use could be determined. The settlers would not participate in the forestry program, and would lose the opportunity to learn how to manage these resources scientifically and how to operate the machinery.

The agricultural program would depend entirely on technology and inputs imported not only from outside the area, but from outside of the tropics. The mechanized cultivation of basic grains such as rice and maize would require the massive application of chemical fertilizers, herbicides, and pesticides. These practices have never yet been viable in the Mexican tropics on a large scale.

The Chinantecs would have no way to apply their traditional knowledge to the program. If the crops failed and the program were to be abandoned, the area would probably be planted with pasture. One of the most complex ecosystems on earth would be replaced with one of the simplest, and there would not be enough work for the resettled population. The ecologists suggested that mechanized agricultural programs should be located in areas cleared in past years, where production had already fallen below profitable levels.

The Commission's Reaction

The Papaloapan Commission's position on the ecological question was articulated in its formal plan for Uxpanapa, which was published in the Mexico City newspaper El Nacional on October 11, 1974 (Tamayo, Rovirosa, 1974). Ing. Jorge L. Tamayo, Executive Director of the Commission, spelled out his ideas about the potential of the Southeast for capital-intensive development in a speech given at the Universidad Veracruzana in Xalapa, Veracruz, on November 29, 1976 (Tamayo, 1976).

The directors of the Papaloapan Commission believed that the ecologists' proposals were totally impractical, and felt personally threatened by the public opposition to their policy. They were responsible for the Chinantecs displaced by the dam and the existing population, or well over 20,000 people. The new Agrarian Reform law requires ejidos to be organized collectively to facilitate financial management, credit, and mechanization. The commission had just found a way out of a serious conflict over the Cerro de Oro project itself, and was well funded for the first time in nearly 20 years. To organize a project as large and as expensive as Uxpanapa, they had to balance the interests of diverse groups and centers of political and economic power in Mexico City, Oaxaca, and Veracruz.

The national food production crisis was the principal argument for opening up the new land. The construction of an extensive infrastructure in a virgin tropical area required capital and heavy machinery, which they could not have obtained unless they could promise a return at least as high as that of major projects in other parts of the country. The cost of developing new agricultural land with irrigation had risen to over three thousand dollars a hectare (Tamayo, 1976). If the capacity to produce the basic grains becoming increasingly short in Mexico could be developed in the tropics at significantly lower cost, the capital that had been invested in the North could be shifted, with multiplying benefits for all sectors in the Southeast.

This could only be accomplished if population and resources were concentrated, and was incompatible with scattered settlement patterns and diverse, small-scale enterprises. The success of this policy depended on the development of technology with higher yields and lower costs of production than had yet been achieved in the tropics, which required the concentration of resources and manpower. In addition, the commission was legally committed to provide housing and a broad range of services to the Chinantecs, which would have been prohibitively expensive if the population were scattered in small communities.

The commission saw the ecologists' proposals as an irresponsible attempt to consciously set up a system of subsistence agriculture, which might or might not slowly develop into more productive systems if totally new and untried avenues of research were successful. Instead of producing the specific crops for which there was an immediate and pressing demand, they would be experimenting with new products which might not find a market. The rain forest, for all its enormous biomass, species variety, and scientific interest, had not provided economic benefits to anyone except the Americans who took out the best trees at the beginning of the century. Large volumes of timber had been wasted in other areas in the process of agricultural development. The commission believed that an organized program of commercialization would minimize the waste of the timber resources, even if it could not realize the theoretical value of each cubic meter of wood.

The environmental issues the commission did take seriously were erosion and silting, problems with which they were very familiar in the Papaloapan basin. Its official response to its critics was as follows (Tamayo, Wade, 1974):

> The utilization of the tropical jungle has been much discussed, both in Mexico and in other tropical countries. In general, experience has been negative, primarily because the methods by which erosion can be avoided and controlled have not been considered, and the areas to be cleared have been poorly chosen. To avoid these errors, an aerial photography study has been made. On this basis, we have decided to clear only lands on a slope of less than ten percent which are appropriate for (mechanized) agriculture. Thirty-three percent of the zone will be cleared; the agricultural areas will be interspersed with forest cover which will protect the soil. Strips of the original forest 20 meters wide will be left every 200 meters, and fast-growing species will be planted in these windbreaks to replace the trees which will be harvested.
>
> We have been in constant consultation with many ecologists, with the National University, and particularly with Dr. Gomez-Pompa so that they may help us to avoid conditions which favor erosion. We have prepared a detailed plan for the clearing operation.
>
> We have set up an experiment station to study the environmental conditions, to test the crops which will be introduced into the zone, and to try to find new or complementary techniques to improve the traditional crops of the area.
>
> In the clearing operations which have been undertaken in other areas in the past, the trees have been burned because their commercial value did not justify their utilization. Because of the large volumes involved in Uxpanapa, we plan to sell as much wood as possible to meet the shortages of commercial timber in Mexico, and especially the railroad tie deficit which the national system is experiencing. The benefits of this operation will be shared with the ejidatarios.

The ecologists took their point of view to President Echeverria and to the governor of Veracruz, but were overridden. They were not invited to return to Uxpanapa and have been able to publish only a few preliminary reports. They made a film for television that was sharply critical of the commission. Ever since, the project has been administered by civil engineers and agronomists. No one trained in tropical biology has been studying the effects of the clearing or has been able to participate in the program on any level. Dr. Gomez-Pompa commented as follows (Gomez-Pompa, 1975, p. 11):

> Uxpanapa . . . will sadly pass into history as a region which could have been incorporated into the national development process in an orderly fashion. Unfortunately, the basic decisions have been made without taking into account the possibility of using the natural resources which exist, or rather existed, in the region.

The project has been severely criticized in the press, both in Veracruz and in Mexico City. An article appearing in the national newsmagazine <u>Proceso</u> in 1977 entitled "Ethnocide and Ecocide" summarized the positions of the anthropologists in Ojitlan and the ecologists in Uxpanapa. In reaction to such opposition, the commission came to distrust outside expertise and to depend almost entirely on its technical staff and that of a limited number of other government agencies. This simplified administration and allowed the Uxpanapa project to proceed according to its original plan. Nevertheless, the isolation from the ideas and experience of others has been unfortunate.

ROADS

The first and most important construction was a temporary access road four to six meters wide running 140 kilometers through the zone from the Trans-Isthmian highway to the valley of the Uxpanapa river. It was started in August, 1973 and was completed in a little less than two years. Lateral roads are of two types: ungraded tracks for the lumber companies at one-kilometer intervals, and graded, all-weather roads at five-kilometer intervals. A permanent road nine meters wide is being built parallel to the original and will be paved as soon as construction is completed and the heavy machinery pulled out. This will eventually connect with a new federal highway that will traverse the Isthmus from Cintalapa, Chiapas, to Sayula, Veracruz. This route will greatly reduce the distance between Mexico City and Central America, and will put Uxpanapa at an important crossroads.

The construction of the access road was pushed forward under great time pressure so that the lumbering, clearing, and other activities could begin as rapidly as possible. A 120-meter Bailey bridge was built across the largest of the rivers in a few months. Including a large airstrip, the costs of the basic communications system took up nearly half of the budget in the first stage.

Eight camps were built to house the engineers, the construction workers, and those Chinantecs who came to build their houses and work their fields before the towns were completed. One of them, La Laguna, is the headquarters and is being expanded into a permanent administrative center for the zone. The agro-industrial complex will be located nearby.

THE TIMBER OPERATION

When the forestry surveys were made, the government was already committed to a policy of clearing large areas of the forest and opening up the land to agriculture. Tropical forestry is a complex business, because the density of any particular species is very low. Lumber, paper, and other technology for the elaboration of wood products first developed in temperate regions around a relatively small number of species with specific and well-understood properties. For many years, commercial activities in tropical Mexico were limited to the extraction of mahogany, cedar, and certain other valuable woods. In isolated areas, the costs of building roads and getting the logs to the nearest river or railhead has meant that selective, sustained-yield timbering has not been an economic proposition. Enormous quantities of valuable timber have been wasted as lands have been cleared for agriculture, both by slash-and-burn cultivators and by planned colonization projects.

There are 20.6 million hectares of tropical forest in the Southeast and along the Pacific coast. The national annual harvest of six million cubic meters of logs falls short of demand by 25 percent (Echenique-Manrique, 1975, p. 12). Inadequate research and marketing systems have prevented the use of many valuable tropical species. For example, the Mexico City Metro mass transportation system was built in the 1960s by a French company. The line carries thousands of people every day, and exacting safety specifications were set for the wooden ties so that they would last for at least 30 years. Tropical hardwoods are extremely durable, so the French used the azobe tree from one of their colonies in Africa when they built the Metro in Paris. When bids were offered in Mexico, no company was capable of providing ties of the necessary quality within the specified time. Wood was imported from Africa, although six suitable species have since been identified in the Mexican tropics (Echenique-Manrique, 1976). The National Railroad has used ties imported from the United States and Canada for many years. Balance of payments problems led to cutbacks that resulted in a net deficit of ten million ties, in spite of large national forest reserves (Tamayo, 1976b).

Timber operations under government auspices are expanding rapidly. Recent developments in paper, plywood, and particle-board technology allow the use of mixed tropical hardwoods, if the economic return is great enough. In many areas of the world, the pendulum has swung from limited high-grading to the opposite extreme of extensive clear-cutting. In Southeast Asia, for example, expanding demand in developed countries has led to the denudation of thousands of hectares without adequate conservation, management, or reforestation programs (Far Eastern Economic Review, 1977; Unasylva, 1976). Tropical hardwoods do not propagate easily and grow very slowly. In some parts of the Mexican Southeast, plantations of fast-growing softwoods such as Caribbean Pine have been established to produce paper pulp (Tamayo, 1976a). These factors add to the pressures on the remaining virgin areas, which may disappear before effective, sustained-yield management systems, which take advantage of a wide variety of species, can be developed.

In Uxpanapa, the government decided to sell as much timber as possible before clearing the land. The lumber operation would both use and help pay for the road network necessary for the successful development of intensive agriculture. A special federal authority was set up to finance and coordinate the activities of fifteen private lumber companies. Skidders, trucks, and other heavy machinery were provided on credit to the contractors, and 17 sawmills were installed. The government's initial investment was about 80 million pesos (Mexico, 7).

The companies have been free to cut whatever trees they want, with preference for the areas on a slope of less than ten percent, which will be cleared for agriculture. Two hundred sixty thousand hectares is a large area, and the national demand for specific types of wood was quickly saturated. The initial studies identified over 100 species of trees, only a few of which have ever been sold commercially or studied in any detail. Of the 325,000 cubic meters which were sold up to 1977, three quarters went out as undifferentiated tropical hardwoods for uses where the specific properties of the wood do not matter. The best single market has been the National Railroad, which took delivery of about 800,000 ties. Only a few mahogany, cedar and other valuable trees were missed by the Americans 75 years ago. It was originally estimated that 50 cubic meters of wood could be sold off each hectare, but pressures of time and problems both in the forest and in the market have kept the actual extraction rate well below this figure (Mexico, 7).

The private contractors have made good profits, and the government is recuperating its capital investment. The ejidos have received an average of 48 pesos, or between two and three American dollars per cubic meter as their share. This is approximately equivalent to .6 American cents per board foot.

CLEARING

According to the original plan based on the initial photo survey, 85,000 of the 260,000 hectares in the zone were to be cleared for agriculture. To minimize the danger of erosion, only those areas on a slope of less than ten percent were to be cleared by machine. Steeper slopes suitable for pastures, rubber, and other perennial crops would gradually be cleared by hand. On the maps, the entire zone was divided into a perfectly regular grid of fields one kilometer by 200 meters, separated by windbreaks oriented perpendicular to the prevailing winds. These rows of trees were intended both to preserve some of the original vegetation and to protect the crops.

As soon as the lumber companies had taken out the valuable timber, the work began. In a tropical forest, most of the nutrients are concentrated in the top layer of the soil. Even very large trees have shallow, horizontal root systems. Two bulldozers dragging a heavy chain between them, assisted by a third machine pushing behind, can mow down the forest with astonishing speed. Only the very largest trees must be cut with a chain saw. Bulldozers equipped with special rake blades dig out the roots and stumps and push the fallen mass into piles. The heaps are doused with kerosene and burned, pushed into new piles, and burned again. Two passes with very heavy disks break up the trash and work it into the soil. Although some sticks and trash remain, the land is ready immediately for the mechanized production of annual crops.

During the first stage of the program, between 1974 and 1976, a little under 10,000 hectares were prepared in this way (table 7.2). This was more than the agricultural program was organized to handle at the time, and the rhythm slowed nearly to a halt during the next two years. The goal for the remaining forest was modified on the basis of a more accurate evaluation of soil quality. Nine percent of the soils cleared in the first stage proved extremely rocky and unsuitable for mechanized agriculture. Thin soils, poor drainage, and other problems limited fertility over large areas. By 1978 it was thought that between 28 and 30 thousand hectares would be cleared by machine by the time the program was completed in the early 1980s, or about a third of the original estimate. This will average about five hectares per family, and will be supplemented by whatever is cleared by hand for pastures and perennial crops.

The clearing process requires specialized heavy equipment, and is very expensive. The cost rose from a first estimate of 5,000 pesos per hectare in 1974 to over 12,000 pesos by 1976. After the peso was devalued in that year, the figure went up to over 23,000, maintaining an exchange value of about 1,000 dollars per hectare.

The timber operation has been organized and financed separately, and the income from sales has not been reported on a per-hectare basis. The gross payments of the lumber contractors have been approximately 30 percent of the cost of the subsequent clearing, only a small fraction of which has been received directly by the commission. These funds have been credited against the cost of the road system, not the clearing.

Gaps left by the harvesting of the large trees and the irregularities of the forest have prevented the windbreaks from functioning effectively. Starting in 1977, the land was completely cleared and the residue was piled up in windrows around the perimeter of the fields. Fast-growing trees such as Melina arborea were planted in them to establish effective windbreaks as quickly as possible. The commission hoped that this procedure would reduce costs to 20,000 pesos per hectare and result in better land preparation.

The land in ejidos belongs to the state and cannot be bought or sold. The market value of similar land in privately-held areas in Veracruz varies between three and five thousand pesos per hectare in isolated areas to 12 to 15 thousand for superior soils near all-weather roads (Mexico, 33). The Papaloapan Commission paid a very high price on the expectation that the investment in mechanized clearing would pay off in the production of annual food crops. Land preparation for pasture, rubber, and other perennials is perfectly adequate if done by hand, which costs less than ten percent as much - between 1,700 and 2,300 pesos per hectare depending on the density of the forest cover and the average size of the trees. This work provides employment to the ejidatarios rather than to outside contractors. Another choice would have been to clear by hand, grow subsistence crops for a few years using traditional slash-and-burn technology, and mechanize as the roots and stumps decomposed. The decision to grow annual crops on a large scale

Table 7.2. Uxpanapa: First Stage of Clearing Operation, 1974-1976

(hectares)

Ejido	Total Area of Ejido	Cleared by October, 1976: Machine Cleared: Suitable for Mechanized Agriculture	Machine Cleared: Not Suitable	Hand Cleared	Total	Planted in 1976: Area	Percent of Cleared
Carolina Anaya*	3,285	21	-	80	101	174**	n.a.
Saturnino Cedillo*	1,278	230	127	66	423	60	14
Benito Juarez I	1,062	578	19	90	687	365	53
Plan de Arroyos*	3,285	418	10	35	463	314	68
Miguel Aleman	750	225	-	15	240	250	100
Francisco Javier Mina	1,312	25	-	100	125	n.a.	n.a.
Monterrey*	1,090	373	4	-	377	290	77
Alvaro Obregon 16	1,792	116	30	-	146	82	56
Augustin Melgar*	1,920	2	23	130	155	125	81
Hermanos Cedillo*	2,786	405	54	100	559	250	45
Almanza	1,160	504	-	15	519	264	51
Murillo Vidal	n.a.	401	27	26	454	221	49
Celestino Gasca	880	119	36	16	171	63	37
Benito Juarez II*	n.a.	158	1	15	174	370**	n.a.
Benito Juarez IV	n.a.	581	18	6	605	132	22
San Antonio	1,330	195	-	8	303	50	17
Benito Juarez V*	n.a.	252	1	63	316	123	40
Buenavista	1,360	271	45	5	321	100	31
El Pilon	610	108	33	52	193	88	46
Adalberto Tejada	1,000	138	13	40	191	79	41
Salta de Eyipantla	5,860	171	15	-	186	62	33
Adolfo Lopez Mateos	2,744	701	120	-	821	495	60
Alvaro Obregon 40	2,450	431	57	-	488	392	80
Josefa Ortiz de Dominguez	1,140	359	63	-	422	95	23
Helio Garcia Alfaro	7,300	1,521	93	10	1,624	1,523	94
TOTAL		8,402	787	872	10,061	n.a.	n.a.

*Ejidos settled by spontaneous colonists.

**Includes areas planted in other ejidos.

n.a. = not available

Source: La Comision del Papaloapan; La Secretaria de la Reforma Agraria.

immediately implied further investments in tractors, implements, storage facilities, and so on.

The soils of Uxpanapa are more fertile than those found in many areas of the tropics, but they present complex management challenges. In spite of efforts to minimize this problem, mechanized clearing does destroy part of the layer of organic matter. The heavy equipment does compact the soil to some extent, in spite of its strong structure. Maize yields have been highest in areas cleared by hand, lower where the machine did the work, and lowest where the bulldozers passed often or turned around. The size of the project and the speed with which it has been brought into production have not allowed important differences in soil quality over short distances to be taken into account.

In 1975, when the operation was in full swing, hundreds of fires were burning at once in clearings all over the forest. It was a very dry year and some of them escaped, causing a serious forest fire which burned for several months and damaged thousands of hectares of valuable timber. This caused some political embarrassment when President Echeverria and Jose Lopez Portillo, then minister of the treasury and currently president, visited the zone that May.

THE NEW TOWNS

One of the most striking differences between Ojitlan and Uxpanapa is the appearance of the towns. The traditional villages are typically clustered around a single main street. The houses, which are scattered irregularly among the garden plots and fruit trees, are built of bamboo with steeply pitched thatched roofs. They are often built as double structures; one for living and sleeping, the other for cooking. This basic design is universal in the Mexican tropics. In Uxpanapa, the population is being concentrated in twelve new towns to facilitate the installation of drinking water, sewage, and electricity.

The commission contracted several studies to investigate the attitudes of the Chinantecs toward housing (Mexico, 19). Architects built models that were exhibited in Ojitlan so that the people could choose what they liked. In what is really the only example of direct Chinantec participation in the Uxpanapa project, the people are building their own houses with a 20,000-peso subsidy, which is provided in the form of building materials and as a daily wage. Architects and engineers have worked alongside of them to help with design and structural problems (Carrion Spindola, Cruz, Cruz, 1977; Escanero, 1977).

In the early 1950s, many of the Mazatecs in the Temazcal resettlement zones refused to modify their traditional housing, even when the government built concrete floors and frames for them. Over the past 20 years, the values of the larger society have penetrated deeply into the Chinantec communities. Modern houses have become a mark of social status. Almost all of the settlers are building nearly identical structures - simple one-room blocks built out of locally produced concrete bricks roofed with corrugated asbestos or zinc sheets. One reason is that the coyol palm used for thatch is not found in the primary forest of Uxpanapa. Another is that the subsidy is distributed in the form of building materials that cannot be substituted for cash. Although a few Chinantecs have spent extra money on larger and more impressive structures, most have tried to keep costs to an absolute minimum.

There have been two important points of difference between the Indians and their technical advisors. The architects have felt very strongly that the houses should be divided into several rooms, primarily so that the parents would be separated from the children at night. This idea has been almost universally rejected. The other has been the location of the bathrooms. The Chinantecs have been pleased to have plumbing, but the idea of installing it inside the house has seemed very unsanitary. They have insisted on a separate structure, or at least a separate entrance from the outside.

The towns are laid out on a network of streets radiating off a central avenue. Each house is provided with piped water and sewage. The commission contracted with the Federal Electricity Commission to install distribution lines to each house. A civic center consists of a primary school, a health clinic, a post office, and a meeting hall.

Sixteen settlements were projected and sited according to the original aerial photo survey. On the maps, it seemed logical to place them strategically at points equidistant from the fields on soils which could not be used for agriculture. This put them in hilly, rocky areas where the costs of construction were very high. In one case, work was begun in the dry season in an area which was seriously flooded a few months later by the discharge from the underground river. A diversion canal had to be built. Many of the garden plots behind the houses are filled with large boulders and cannot be used for trees, vegetables, or crops.

Six of the eight settlements of the first stage are located far from any river or stream. The Chinantecs have always lived very close to the rivers in their homeland, where they spend a great deal of time swimming, washing, and traveling in their canoes. They found these inland locations an unnecessary hardship, especially in the first years before the water systems were completed. In 1977 the commission modified the plan and decided to build four large towns rather than eight small ones to complete the program. These will be located on the banks of rivers on better soils.

8 The First Years of the Agricultural Program

The Papaloapan Commission is introducing modern, highly-mechanized technology into Uxpanapa as rapidly as possible. Over the past ten years, the federal government has initiated similar projects in the Chontalpa and Balancan-Tenosique regions in the state of Tabasco, and in other areas of the Southeast (Chao Ebergenyi, n.d.; Mexico 3). Nevertheless, there is very little experimental or practical experience to guide the development of intensive agriculture in the Mexican rainy tropics. A former director of the agricultural development bureau of the commission has stated the problem succinctly (Rodriguez Vallejo, 1977b).

> . . . agricultural systems and developmental patterns from the temperature regions should not be used to develop the rainy tropics. This error continues to be repeated over and over again in all of the tropical regions of the country. We must develop new technologies, new ways of thinking, and new agro-systems to be able to use the great potential of these areas for the benefit of the local population and the national economy.

Basic and applied research in tropical agriculture has been very limited in Mexico. As in most parts of the world, money has been available to study profitable commercial crops in areas where they are grown intensively. Most of the research facilities and trained personnel are located in the irrigation districts in the Northwest and in Mexico City. In recent years, the government has established new research centers in the tropics, including an agricultural college in Cardenas, Tabasco, but technology applicable to a broad range of natural conditions takes a long time to develop. Several years of careful study before the Uxpanapa project was started would have been ideal, but the schedule of the resettlement program in particular, and the pressures on the resources of the Southeast in general, have forced the Commission to move rapidly.

EARLY EMPHASIS ON ANNUAL GRAIN CROPS

From the day the project got underway, the managers have been faced with a dilemma. The national production crisis created political pressures that required them to grow annual grain crops, specifically rice and maize, for which there was a pressing need. This forced them to use the available technology, even though it had not been tested in the rainy tropics on a large scale. At the same time, the long-term success of Uxpanapa depends on the development of economically viable, sustainable systems in which the area will have a relative advantage. Cattle operations and perennial crops, especially natural rubber, are intended to provide the ejidatarios with an improved standard of living and to transform the forest into a center for regional growth.

Test plots of maize, rice, and a few other annuals were set out when the program first began to determine very basic information, such as fertilizer response. As soon as field operations got underway on a large scale, the commission's small staff put most of its emphasis on the day-to-day problems. INIA, the national research institute, ran a few simple experiments as part of a series of regional variety trials. The results were processed so slowly that they were not available in Uxpanapa in time to affect the plan for subsequent seasons. The agronomists chose the varieties and practices they thought held the best probability of success, and have been modifying the systems on the basis of experience.

The principal goal of the program in the first years was to plant the entire cleared area with annual crops in both seasons. This land, which totaled about 10,000 hectares, was located in an irregular strip along 100 kilometers of dirt road, which complicated the efficient allocation of the agricultural machinery and other inputs. The movement of the Chinantecs into the zone was slow, so that the area cleared in each ejidos did not correspond to the number of people who were available to work. These problems were compounded by unpredictable weather conditions.

In the first four years, between 30 and 80 percent of the cleared area was left unsown. Secondary growth is extremely rapid, and requires expensive reclearing within a few years. The Papaloapan Commission was unwilling to bring in workers from the outside, for fear that they would not leave when the Chinantecs arrived. The spontaneous settlers were offered short-term credit contracts to work the cleared areas, but most of them preferred to grow crops on their own land using traditional slash-and-burn technology. In 1978, it was estimated that 3,060 man-days of labor would be required to cultivate all of the cleared land in the summer season, of which only 1,750, or 57 percent, were even theoretically available from the resident population (Mexico, 13).

In the first seven seasons, credit was granted primarily to ejidos working with the newly introduced mechanized systems on land cleared by machine, but some slash-and-burn cultivators also received assistance. The results are summarized in table 8.1.

The yields were extremely low. The averages include large areas that were never harvested at all. Some fields of rice produced over four tons per hectare; maize yields of 2½-3 tons were reported. During this same period, the spontaneous settlers increased the area under cultivation using traditional methods without credits in response to improved communications with the markets. They chose their plots with care and did not plant more than they could harvest. Their yields were significantly better.

With the exception of one small area of chili, the average yields of all of the annual crops grown with the newly introduced mechanized systems were too low to recover the costs of production. The ejidos harvested part of their crops, but none of them made any profit. The actual cash losses recovered from crop insurance, absorbed by the commission, or charged against the ejidos as bad debts, have never been officially tabulated and released. The accounting system of the bank is tied to credit allowances issued at the beginning of each season, so that the actual costs can only be imputed indirectly. The following credit balances, based on unofficial data, give some idea of the short-term losses which the federal government was willing to absorb in its attempt to introduce annual crops into Uxpanapa (Mexico, 1):

Season	Area Contracted (hectares)	Credit Allowances (millions of pesos)	Value of Sales (millions of pesos)	Recuperation Rate (percent)
Summer 1976	5,620	23.7	5.2	22
Summer 1977	4,250	16.0	5.9	30

Although the recuperation rates were very low, they were higher than the average for official credit in tropical lowland areas (Rodriguez Vallejo, 1977b). Production has been limited by a complex combination of climatic, biological, institutional, and organizational factors. The systems that are being introduced, and the nature of the problems, will be explained in detail in the following sections.

THE INSTITUTIONAL STRUCTURE

Both the Echeverria (1970-1976) and the Lopez Portillo (1976-1982) administrations placed a great deal of emphasis on the reorganization of the bureaucracy serving the agricultural sector. The institutional structure being set up in Uxpanapa reflects new laws and procedures intended to revitalize the ejidos and give them a key role in the

Table 8.1. Uxpanapa: Annual Crop Production with Credit, 1975-1978 (area in hectares; yield in kilograms per hectare)

Season (Number of Ejidos Participating)	Crops	Area (hectares)	Average Yield (kgs/hectare)
Winter 1974-75 (5)	Maize--Hybrid (H-503) and Native	340	200
	Beans--Black	180	180
	TOTAL	520	
Summer 1975 (5)	Maize--Hybrid (H-507) and Native	700	700
	Rice--Dwarf (Sinaloa A-68)	810	negligible
	TOTAL	1,510	
Winter 1975-76 (11)	Maize--Hybrid (H-507)	2,400	800
Summer 1976 (23)	Maize--Hybrid (H-507) and Native	2,420	700
	Rice--Bluebonnet-50	2,000	1,000
	Rice--CICA-4	1,200	negligible
	TOTAL	5,620	
Winter 1976-77 (23)	Maize--All Native Varieties	1,500	600
	Chili	140	5,000
	TOTAL	1,640	
Summer 1977 (39)	Maize--All Native Varieties	3,200	485
	Rice--Bluebonnet-50	1,050	475
	TOTAL	4,250	
Winter 1977-78 (31)	Maize--All Native Varieties	2,000	Not Available

Source: Areas planted as reported by the Banco de Credito Rural del Golfo. The average yields were calculated by the author from unofficial estimates of total sales in each season.

modernization of the rain-fed areas of Mexico. This means that both a new technology and a newly reorganized type of economic and social organization are being introduced into Uxpanapa at the same time.

THE COLLECTIVE EJIDOS

The ejidos in Ojitlan were rural communities of families, which worked a small parcel individually. The elected authorities took care of issues that affected the group as a whole and provided a link with the local political structure. Each campesino decided which crops he would plant, arranged with the local merchant for credit, and drew upon his relatives and friends for labor in peak periods. Official credit was provided sporadically to some of the Chinantecs, who were usually organized into "Solidarity Groups" of three to fifteen members who pooled their responsibility to repay the loans. Except for coordinating the use of common woodlots, pastures, and the plot that supported the school, the ejido as such had no direct economic function.

The division of the ejidos into tiny minifundia has become a serious barrier to agricultural modernization throughout the country. During the Echeverria administration, the government rewrote the basic laws in an attempt to establish the ejido itself as the fundamental economic unit in the reformed sector. The history of the collective ejidos created in the 1930s has been outlined in Chapter 2. After many years of neglect, this type of production cooperative was revived as a way to combine individual plots into units large enough to use tractors and other modern inputs efficiently. Economic and personal interests in established communities have frustrated the attempt to collectivize existing ejidos. The official bank has organized many new credit associations, but it has not been able to change the patterns of control over productive resources that have evolved over several generations.

As Uxpanapa was opened up for the first time under direct federal jurisdiction, the new agrarian reform and credit laws required the organization of collective ejidos. The theoretical advantage of the system is that it combines democratic participation by the campesinos with the centralized technical direction the development of efficient, large-scale agriculture requires. All of the land is held in common and each ejido is organized as a kind of corporation that manages the various enterprises in which it is engaged. Its operations are supervised by the Ministry of Agrarian Reform, which is responsible for its legal structure, and by the official bank, which provides the working capital.

The ultimate authority is the general assembly of legal ejidatarios: all men over 16 years old on the day the organization is formed, and women who hold rights they have inherited. The body is responsible for approving all technical plans and credit contracts, electing official representatives, and any other business that may be proposed. At the end of each cropping cycle, a special planning and evaluating meeting is held at which the bank and other government institutions are represented. The profits from the harvest are distributed, usually according to the following formula: 30 percent in equal parts to all of the ejidatarios, and 70 percent in proportion to the number of days each member has worked, at a uniform daily rate that is independent of productivity. At the same time, the technical plan for the next season is approved according to the government's plan for the region, the availability of credit, and the wishes of the members.

The executive body, or Comisariado Ejidal, is composed of a president, a secretary, and a treasurer. Its term runs for three years. The president has very wide powers. He presides at the assembly and draws up the agenda. He represents the ejido in all of its relationships with outside authorities, particularly the official bank, and directs the day-to-day operations of the farm. Credit is not received in a lump sum, but rather in the actual inputs; he must sign a receipt for each bag of seed, fertilizer, and so on, as they are delivered. He organizes the daily work schedule and distributes the advance payments made to the members. He writes an endless series of petitions to various government agencies to resolve conflicts and receive services. His work involves frequent trips to local offices, the state capital, and even to Mexico City. Although it is a full-time job, he does not usually receive payment for his work. He delegates a member of his family or hires someone to do his share of the agricultural labor. Presidents are sometimes accused of appropriating funds for their own use. For obvious reasons, they must be literate in Spanish, and come from the stratum of Chinantec society which is the most fully acculturated into the larger society.

A second executive body, the Vigilance Committee, or Consejo de Vigilancia, is headed by the man who comes in second in the election. He is responsible for keeping an eye on the president, keeping track of the accounts, and countersigning the credit receipts. His actual authority varies considerably from ejido to ejido.

The Union of Ejidos

Following the terms of the decree that authorized the construction of the Cerro de Oro dam, each individual in the lakebed was paid indemnification for the value of his personal property. The value of the land and the other common assets of the communities was paid into a fund to be invested in Uxpanapa for works of mutual benefit. This was supplemented by the income from the lumber operation, and totaled 150 million pesos. Rather than pay the money to each new ejido, a Union of Ejidos was organized to administer it. As the name implies, this is an association of ejidos, each one of which elects representatives. Its first investment was a central machinery station, which was set up in 1978 under the direction of a professional manager. Twenty percent of the 83 million peso cost came from the fund; 80 percent was provided as credit from the official bank.

Future plans for an agro-industrial center include warehouses, a gasoline station, a rice mill, and a bread factory. In general, the union will assume the functions usually performed by private merchants and entrepreneurs. It is hoped that the surplus generated by the primary activities in the zone will be reinvested locally. This is meant to provide employment for the sons and daughters of the ejidatarios and to provide the basis for balanced regional growth.

Many of the Chinantecs, and nearly all of the spontaneous settlers, have been suspicious of this institution. Believing that the common funds belong to them, they have been frustrated that the Union has not allowed them to invest in trucks, tractors, cattle, fencing, and other needs in their own ejidos. They fear that the union will be, in effect, another agency of the government, over which they will have little effective control.

The Agrarian Reform Ministry

The Agrarian Reform Ministry (the Department of Agrarian Affairs and Colonization (DAAC) was given ministerial status during the Echeverria administration) has very broad responsibilities over the legal titles and formal organization of the ejidos. It is plagued by byzantine bureaucratic procedures, and simple business is often dragged out over many years. All of the employees of the local office were fired in 1975, when they were charged with drawing the ejidal boundaries so as to exclude a large area of good land they planned to sell for profit. For three years after this incident, while the agricultural program was first being organized, the ministry had no direct representatives in Uxpanapa at all. They are supposed to explain the cooperatives, the laws, and the credit system to the ejidatarios and work out the details of their organization with them. This vital function was taken over by the commission and the bank, whose overriding interest was to get the cleared land planted as quickly as possible.

When Uxpanapa was first designated for resettlement, a census was taken of the spontaneously-settled ejidos to determine the number of vacancies available for the Chinantecs, and the area was surveyed. A variety of conflicts with the spontaneous settlers arose over the prior documentation of their boundaries. Another problem was that campesinos who were not counted in the census for one reason or another presented evidence that they had legal rights to live in the area. The ministry and the commission looked at these petitions on a case-by-case basis, and accepted some of them. By 1976 they had recognized the prior rights of over 1500 ejidatarios, considerably more than the number of Chinantecs who had moved in by that time. To protect the land for the resettlement program, the government has guarded the zone against invasion from other areas. The perimeter has been patrolled by a special police force and a symbolic detachment of the army, and a pass has been required to enter.

The Rural Credit Bank

The Banco de Credito Rural del Golfo was formed in 1975 by the merger of the Ejidal Bank with two other official credit institutions. The local office has played an important and difficult role in Uxpanapa. Not only does it finance and supervise the inputs and labor used in each ejido; it is responsible for organizing the flow of tractors, crop-spraying planes, seeds, chemical products, and other factors of production into the zone and their efficient allocation. It markets the harvests, distributes the profits, and is the only source of income for most of the ejidatarios. Credit can only be granted on a collective basis, so the bank's inspectors must deal with the complex problems of ejidal organization every day. Its staff is small and it must process an enormous quantity of paperwork.

There are two types of credit. Production credit, or credito de avio, covers the costs of production of annual crops and is closely supervised on a hectare-by-hectare basis. The inputs and the cash advances the ejidatarios receive for their labor are paid as each task is completed. The Bank sells the product, deducts all costs plus interest at an annual rate of 12 percent, and distributes the residual among the ejidatarios.

Investments in cattle ranches, rubber plantations, and other enterprises with a large initial cost and a slow rate of recuperation are financed through long-term credit, or credito refaccionario. The capital is provided by the central Bank of Mexico through an organization called FIRA, but the local office of the bank is responsible for day-to-day operations. The funding agencies require very careful technical and financial studies, so the approval process is slow. The repayment period is between 12 and 15 years and the interest rate, after a two-year grace period, is 9 percent.

The first step in each agricultural cycle is the choice of crops. Acreage targets are set at a meeting of the regional rain-fed agriculture district in Coatzacoalcos by representatives from the bank, the Papaloapan Commission, and other regional and national agencies. Lines of credit are approved for several crops, according to national priorities, the availability of markets, and local conditions. In cooperation with the crop insurance company, whose functions are outlined in the next section, the regional office of the bank decides how much credit will be allowed for each crop on a per hectare basis, on several levels of technology from slash-and-burn to fully mechanized systems. The total is then divided up into separate operations tied to a fixed calendar.

These estimates of the cost of production are important documents for several reasons. The cash advances the ejidatarios receive in the course of the season are not a salary; they are part of the credit and are deducted from the final profits. If, for example, the seeding operation is allocated 250 pesos per hectare, each ejidatario receives 250 divided by the number of people required to do the work. A highly mechanized system is being introduced into Uxpanapa; the labor allowed by the credit varies between crops and levels of technology, but it is relatively little. If the advance payments are averaged over the number of working days in a cycle, they come out to as little as ten or twenty pesos a day per person. This is less than the Chinantecs received in Ojitlan for working in their neighbors' fields, and is well below the federal minimum wage the Papaloapan Commission is required to pay the people it hires in the zone (the minimum wage is tied to the cost of living in each area of the country; in Uxpanapa, it inflated from 56 to 101 pesos/day between 1976 and 1978).

The allowances also detail the costs of the various inputs. These are supplied directly by the bank, so the values listed would simply be an internal accounting measure if the needs could be accurately estimated in advance. In fact, serious attacks by pests and diseases have required heavy and repeated applications of expensive chemicals. The bank must maintain the appearance of staying within the predetermined costs to meet the requirements of the insurance company and for other accounting purposes. To prevent the loss of the crop, it "borrows" from one operation to pay unexpected over-runs in another. If this is not enough, the local branch requests permission from the regional office for an overdraft. Although the ejidal presidents must approve this, the theoretical costs of production rapidly become a fiction, and the ejidatarios lose control over what is spent on their land.

Meetings are held in each of the ejidos at the beginning of each cropping cycle. The members are told what is available and are asked to decide how many hectares of each crop they would like to grow. The government is under a great deal of pressure to plant as much of the cleared land as possible, and the ejidatarios have little effective choice. A credit contract is prepared for each crop, which must be approved by the majority if the operation is to proceed. In some cases, agents of the Bank have been sent to Ojitlan to collect signatures from the people who have not yet moved.

Inspectors from the bank share responsibility for the supervision of daily operations during the season with the presidents of the ejidos and the agents of the Papaloapan Commission. They are not agronomists, and their job is to make sure that the work is performed according to the contracts. Although they do not formally provide technical assistance, they are in the field every day and play a key role in the introduction of the new technology. They make the process as flexible as they can, but they must meet the rigid requirements of the system, at least on paper. The bank retains ownership of the product until it is sold.

The Crop Insurance Company

In the 1940s, growers of crops with very high costs of production, such as cotton, formed mutual insurance associations in the irrigation districts in northern Mexico. Each member paid a fixed quota per hectare and the money was distributed among those farmers affected by some natural disaster. In 1961, all such associations were consolidated into a nation-wide federal agency, the National Crop and Cattle Insurance Company (ANAGSA). The organization is designed to complement the services provided by the official bank by insuring its capital stock against the inability of the farmers to repay their loans in bad years, and is a prerequisite for credit (Mexico, 34). In Uxpanapa it plays a very important role by protecting the investments of the bank, but it does not guarantee a profit to the ejidatarios.

A complex set of requirements and procedures, which are intended to minimize the company's risks and to avoid fraud, introduce irrational rigidities into the management of the crops in each ejido. The uniform credit allowances for the region are set jointly by the insurance company and the bank. In addition to the allowable expenses in each category, they specify seed varieties, planting densities, fertilization levels, and a calendar that sets the dates by which key operations such as planting and harvesting must be accomplished. Premiums are collected on a per hectare basis and the rates, as a percentage of the credit contracts, are high:

Year	Maize (mechanized)	Maize (slash-and-burn)	Rice (mechanized)	Rice (slash-and-burn)
1976	5.5%	-	4.6%	-
1977	12.6%	8.0%	6.3%	-
1978	14.5%	14.5%	9.2%	9.2%

Three separate inspections of each field are necessary to validate a claim. The first is made just after the plants emerge to certify that the crop has been planted and that the population is adequate. Any damage must be reported within 72 hours and must be personally certified by an agent. The probable production and the amount of indemnification are set at a third visit just before the harvest begins. Although these procedures are much simpler in the compact fields of a collective ejido than in scattered individual plots, there have been many problems getting the inspections made on time.

SESA

Until March, 1978, when the Union of Ejidos bought 100 tractors, all of the agricultural machinery in Uxpanapa was contracted directly by the bank. Apart from a few private operators with a few tractors each and the crop-spraying planes, this service has been provided by SESA (Servicos Ejidales, S.A.), a subsidiary of the official bank. Uniform charges per hectare cover the direct operating costs, but amortization, transportation, and maintenance are subsidized. The contract system allows the accumulation of experience before the capital is committed to buy. Each ejido pays only for the services it receives, and is not responsible for the risks. Nevertheless, SESA has been blamed for many of the problems that have limited yields in the annual crops program.

The regional office is located in Tierra Blanca, over 300 kilometers away, and its equipment is spread over a four-state area. Most of it is Allis-Chalmers, which is not widely distributed in Mexico. It has been difficult to maintain a large enough inventory of spare parts to meet every contingency under climatic conditions which are very hard on machinery. Many of the operators are not well trained, especially in the operation of complex equipment such as combine harvesters. They are neither capable nor equipped to make even the simplest repairs. One central shop and a few repair trucks have not been able to stay ahead of maintenance requirements, and a large proportion of the machinery is down at any given time.

There has never been enough equipment in the right place at the right time, and it has not been possible to prepare all of the cleared land before the planting deadlines for the crops. The operators are under constant pressure to work quickly and are paid a bonus for each hectare they complete. The quality of the land preparation - which is critical for the effective control of erosion, weeds, pests, and diseases - has been very poor. Mechanical harvesting has run into very serious problems, both because of unsuitable machinery and because there have not been enough combines in working condition to cover the area that has been planted.

The permanent central machinery station established under the auspices of the Union of Ejidos should be able to profit from past experience. It is relatively easy to get parts for the John Deere and Caterpillar equipment they have purchased. Nevertheless, the cooperative management of a new enterprise is almost certain to create new tensions and problems.

The Papaloapan Commission

The agricultural office of the commission provides the overall planning framework, coordinates the activities of the various institutions and administers the experimental programs. It has divided the area into five zones. In each, an agronomist and several assistants oversee the daily progress of the work and cultivate a demonstration plot. A team of social workers visits with the women in the towns to discuss health, nutrition, and child care. A pilot program has been set up to demonstrate how the 1,600 square meter garden plots can be used for intensive food production.

RICE

Production Systems in Mexico

The rice program in Uxpanapa illustrates the problems involved in transferring technology into the rainy tropics. As part of its program to develop the capacity of the Southeast, the federal government is introducing mechanized upland rice on a large scale. Mexico is self-sufficient in this grain, which is a relatively minor component of the national diet. Per capita consumption is only nine kilograms per year, compared to nearly 20 in Colombia and as much as 150 in parts of the Far East (CIAT, 1977b, pp. 10-13; Grist, 1975, p. 498). Approximately 750,000 tons have been produced in recent years, most of it under irrigation (table 8.2).

About half of the national crop is grown on 50,000 irrigated hectares in the northern state of Sinaloa. Experienced farmers, an active program of experimentation and extension, and an established infrastructure, including 45 percent of Mexico's milling capacity, have concentrated rice in this area. Although experimental plots routinely produce the equivalent of six to seven tons per hectare (Hernandez Aragon, 1976, Chart 3) commercial yields average just over four. This compares unfavorably with irrigated yields of over five tons per hectare in Peru and Colombia (CIAT, 1977b, pp. 13-16). Relative to soybeans, safflower, and the other crops that are grown in the same region, rice brings a relatively low price and requires a great deal of water. The investments necessary to intensify production have not been made.

Faced with chronic water shortages in the Northwest, the government hopes to reduce the rice allotments in the irrigation districts and transfer as much production as possible into the tropics, where there is sufficient natural rainfall. Although these areas already account for well over half of the national area planted with rice, most of it is in small plots grown using traditional technology and varieties, and the yields are low. Uxpanapa is one of several projects in the Southeast where mechanized upland rice is being introduced on a major scale. The others include the Chontalpa in Tabasco, the Edzna valley in Campeche, Tierra Blanca in Veracruz, and various parts of Chiapas. This policy runs counter to that of most countries in Latin America, which are increasing production through intensive irrigation and the introduction of high-yielding varieties (CIAT, 1977b; Scobie, Posada, 1977).

Improved Varieties

Among important grain crops, rice is uniquely suited to irrigated production because it does not take up oxygen from the soil and actually grows best in standing water. The flooded field, or "paddy," system has developed over many centuries in Asia to take advantage of this fact. In the past 20 years, there has been an intensive international research

Table 8.2. Mexico: Systems of Rice Production, 1976
(area in hectares; yield in tons per hectare)

System	Area	Percent of Total Area	Average Yield
	(hectares)		(tons/hectare)
Irrigated and Transplanted	25,000	10	4-5
Modified Asian system, in Morelos and other parts of central Mexico			
Direct Seeded with Irrigation	62,500	25	3-4.5
Mechanized production in Sinaloa and other irrigation districts			
Upland	162,500	65	1-2.5
Primarily in the tropics, including the Southeast--both mechanized and slash-and-burn			
TOTAL	250,000	100	2.9

Source: Leonardo Hernandez Aragon, "Arroz" (Mexico: Secretaria de Agricultura y Ganaderia, INIA, Programa de Mejoramiento de Arroz, mimeo, 1976).

effort to increase world production by improving the rice plant. Dwarf varieties have been developed that can be heavily fertilized with nitrogen without lodging, or falling over into the water. In part because most of the world's rice is produced under irrigation in Asia, and in part because the package of inputs necessary for high yields is more easily managed where water can be carefully controlled, very little emphasis has been placed on the breeding of varieties suitable for upland conditions (IRRI, 1975).

High-yielding varieties have been brought into Mexico from the Philippines, Colombia, and the United States. INIA, the national research institute, has used this genetic material as the basis for a series of dwarf rices. These are now used almost exclusively in the irrigated areas and are exported to Cuba and Central America. They have not replaced tall varieties under upland conditions, because they do not compete well with weeds, because they are difficult to harvest by hand, and because they are susceptible to short periods of drought and to a variety of pests and diseases, particularly the blast fungus (Lopez, 1976). A few improved varieties of intermediate stature have been released, but they are adapted only to a narrow range of upland conditions (Marquez, 1976). Native varieties, which consistently produce modest yields at low cost, are planted in most areas of the Mexican rainy tropics.

The traditional rice farmers in the Southeast work on a small scale. They look for heavy, alluvial soils and plant vigorous, tall, fast-growing varieties that are able to out-compete the weeds and survive drought and other stresses in most years. The seed is planted by hand in widely spaced, compact clumps, which facilitates both weeding and harvesting. The plants in each clump support each other against lodging, and the circulation of air in the spaces in-between reduces the humidity of the microclimate. Blast and other diseases are a problem, but the campesinos are willing to accept yields below two tons per hectare.

1976 - Three Thousand Hectares

The first commercial planting in Uxpanapa was made in 1975, when 810 hectares of a dwarf variety from Sinaloa (Sinaloa A-68) were sown. The rice was seriously affected by Blast, key inputs did not arrive on time, there were not enough combine harvesters, and the crop was entirely lost. The next year, both the Papaloapan Commission and the bank made a tremendous effort to bring all of the necessary elements of their system together. With some crucial exceptions, they did remarkably well organizing a complex operation for the first time in an isolated area. Nevertheless, the average yield was extremely low.

Two varieties were planted: the Colombian dwarf CICA-4 on 1,200 hectares and the tall American Bluebonnet-50 on 2,000, in a total of 23 ejidos. The former was developed by an international team of breeders at CIAT-ICA, primarily for use under irrigation (CIAT - Centro Internacional de Agricultura Tropical - is an international agricultural research center in Cali, Colombia; ICA - Instituto Colombiano Agropecuario - is the Colombian national research institute. The two institutions have worked together to adapt improved genetic material from Asia to Latin American conditions and have released a series of new varieties. Their impact has been analyzed by Scobie and Posada, 1977). Although an attempt was made to incorporate resistance to Blast into its genetic structure, it has proved highly susceptible under a variety of experimental and field conditions (CIAT, 1977a, p. F-11; Mexico, 10). It has a short cycle of less than 110 days and must be harvested immediately or it "shatters," that is spontaneously drops its grain. It competes very unfavorably with weeds under upland conditions, and its growth characteristics make it very difficult to harvest. The fact that is was chosen at all was the result of a breakdown in communications between the Commission, INIA, and the national seed company (PRONASE). Bluebonnet-50 has been grown under both irrigated and upland conditions in various parts of the world for many years. It is tall and vigorous.

In spite of numerous organizational problems, the fields were planted successfully. Virgin soil is relatively free from pathogens that affect crops imported from other areas. Most of the fields had been cleared for the first time only a few months before, so there were few weeds even though the seedbeds were prepared hurriedly. Between 90 and 115 kilos of seed per hectare were broadcast from both tractors and airplanes, producing a dense stand that increased the humidity in the microclimate under the canopy, favoring the spread of Blast.

A few isolated tests and field trials revealed a phosphorus deficiency in the soil, so 90 kilos of P_2O_5 were applied per hectare. Nitrogen probably would have been released in sufficient quantity from the freshly cleared land, but the commission decided to apply 60 kilos of N per hectare to speed decomposition and to insure an adequate supply to the crop. The nearly constant cloud cover shaded the rice from direct sunlight, which reinforced the effect of the nitrogen to encourage luxuriant vegetative growth. The Bluebonnet variety, which normally reaches a height of 155-165 centimeters, grew as tall as two meters.

Rice Blast

The most serious single problem was Rice Blast (Pyricularia orzae). Although this fungus attacks rice grown under all systems, it is a serious limiting factor in upland production all over the world. Consistently high temperatures, high stand density, high humidity in the microclimate immediately surrounding the plant, and light, well-drained soils favor its propagation. The rapid uptake of nitrogen by the plant produces succulent growth that reduces the silica concentration in the cell walls and reduces their resistance to the disease. There is a high correlation between nitrogen fertilizer application and Blast infestation. The management problems are complex, because the same practices that lead to vigorous plants and high yield potential are also favorable to the disease (CIAT, 1975; Cheaney, Jennings, 1975; de Alcantara, 1977; IRRI, 1975; Ou, 1972; Thurston, 1975).

In Uxpanapa, the commission relied heavily on the application of fungicides. Chemical control need not be 100 percent effective to maintain adequate yields, but the sprays are very expensive. The value of the grain that would have been lost must exceed the cost of treatment by a significant margin if the economic threshold is to fall within an acceptable range. Fumigation should be combined with accurate weather forecasting and careful monitoring of the microclimatic conditions. Blast is chemically controlled in large areas of Japan, a developed country with a well-established extension system, high average yields, and a very high support price for rice (IRRI, 1965). It is very difficult to keep costs within reasonable limits when introducing the crop into a frontier area.

Uxpanapa is far from the nearest important rice-producing area, and the fields were distributed in a strip along 70 kilometers of road. Nevertheless, Blast spores found the plants and attacked very seriously. The combination of daily rainfall, high humidity, night-time fog, a continuous plant canopy, very permeable soils, and high levels of available nitrogen created almost ideal conditions for the propagation of the fungus. The technical staff was small and overworked. It was simply not possible for the agronomists to find each infestation immediately, much less predict them by monitoring the microclimate in 3,400 hectares. Whenever they located an attack, they treated it with aerial applications of the fungicide benlate, mixed with sevin to control fall army worms and other pests. Benlate is primarily a protective chemical, and its curative effects are limited. As many as six applications, at a cost of 310 pesos per hectare each, were necessary in many areas.

It soon became apparent that the credit contracts had grossly underestimated the costs of pest and disease control. The staff was under a great deal of pressure to make Uxpanapa produce rice. They were unwilling to consider writing the crop off to the insurance, even after it had become clear that the value of production was not going to be able to pay back the credit. They saw themselves fighting a skirmish in the conquest of the jungle, which they were determined to win.

A common example of their dilemma would involve a field that had been sprayed three times and for which the credit allowance had been exhausted. If one of the field agronomists reported a new infestation, the bank manager had to decide on the spot whether he would let it go and collect on the insurance, or spray again and try to save the crop. It might be possible to average the extra cost over all the ejido's fields and stay within the limit, or it might be possible to arrange for an insured overdraft. Given the pressure on all sides to produce rice, he usually gave his authorization. The presidents of the ejidos, afraid that there would be no more cash advances to their people and no chance of a profit if the fields were abandoned, signed the forms.

The CICA-4 variety was almost entirely destroyed by Blast; the little that survived could not be harvested before it shattered. The more vigorous Bluebonnet-50 did respond to chemical control, but the disease caused severe damage. Helminthisporium, insects, rats, birds, and other pests all took their toll. The crop went into the harvest with a very high level of cost already expended, and with its yield potential significantly reduced.

Relatively few ejidatarios had moved into Uxpanapa at this time, and this was their first experience with mechanized agriculture. Twenty hectares had been planted per person, but nobody was given responsibility for a particular area, and they were not trained to detect Blast or insect attacks scientifically. They were intimidated by the airplanes and the other expensive factors of production over which they had no control. Aside from helping to mix the highly toxic chemicals, they had very little to do and made very little in cash advances.

The Harvest - Combines and Hired Labor

Just as the harvest was about to begin in late September, between 30 and 40 percent of the surviving crop was blown down by high winds. It lodged irregularly, so that almost every field was a patchwork of standing and fallen rice impossible to harvest efficiently by machine. Bluebonnet-50 does not shatter easily; the grain from fallen panicles could be gathered in if it did not lie on the ground for so long that it sprouted.

SESA brought 35 small Allis-Chalmers combines into the zone in the expectation that they would be able to harvest 50 hectares a day. They were very poorly maintained; they had not been greased, their belts and chains were loose, their mechanisms were clogged with rotten sorghum left over from the previous season. They broke down constantly. Even if they were in perfect condition, these machines were not suited to the very difficult conditions in Uxpanapa. Rice is routinely combined when it is grown under irrigation. The fields are drained a month before the harvest, which gives the plants time to dry out. A chemical dessicant is often applied to further reduce the straw to absolutely dry chaff. Gangs of combines work down the leveled fields, separating the grain by a system of shakers and fans, unloading into the waiting trucks every few passes.

In Uxpanapa, most of the fields were hilly and uneven. It rained almost every day throughout the harvest season. The plants had grown very tall and had put out a thick mass of watery vegetation. On any given day, the machines had to wait several hours after the rain let up to enter the fields. They had to maneuver around the lodged patches and cut what amounted to thick, wet grass as much as six feet high. Much of the grain stuck to the straw, did not fall down through the separators, and was thrown out the back as trash. A stick left over from the clearing operation would shut a combine down for half a day or more.

The harvest lasted for two months. Between 11 and 23 machines were in working condition on those days when it was dry enough to enter the fields. They concentrated on the easiest areas and left a very high proportion of the grain in the field. Time was lost as the scattered combines drove back to the reception centers to unload. The bank could not find enough trucks to ship the rice in bulk to its own mill in Acayucan. The private merchant who was contracted to buy the grain in the field required that it be bagged by hand. Cutting an average of 500 kilos per day, the machines were able to harvest just under a thousand tons.

Once it became clear that the combines were not going to be able to do more than salvage a small proportion of the rice, the commission attempted to bring in hired labor to cut the lodged patches by hand. There are millions of campesinos in Mexico who work off their own farms or who are entirely landless. The sugar mills in the tropics contract thousands of cutters every year for extremely hard and unpleasant labor. Nevertheless, it is almost impossible to organize a group of workers on a short-term basis without paying them well.

The Chinantec ejidatarios were in a rebellious mood. They had been promised high incomes from a system of modern agriculture in which they had no effective participation and provided them with very little work. They had signed millions of pesos worth of credit receipts the reduced production could not possibly pay. If they had known that they were going to have to cut the grain by hand, they would have planted small areas using their traditional system, which they knew how to manage. Rice that has been broadcast and then lodged falls into a thick, tangled mat that is almost impossible to cut with a machete, and is extremely slow work with a sickle. A man might possibly harvest a hectare, but certainly not 20.

A total of 575 workers were found; a third of them from Ojitlan and the rest from surrounding areas. The commission paid 58,000 pesos to bring them in, but they harvested very little rice (Mexico, 25). The bank faced a difficult dilemma: beyond a certain point, the more it paid the more it lost. It first offered 50 centavos for each kilo cut and threshed into barrels by hand. The insurance company modified its standard procedure and agreed to pay indemnification for everything that was not actually harvested, which took the burden of increasing costs off the ejidatarios. The commission agreed to make some payments in lieu of profits. The imported campesinos refused to work for less than the minimum wage, 56 pesos a day, which they could not make at the rate they were offered. They were dissatisfied with the high cost of food, which absorbed most of their earnings, the poor accommodations, and the general disorganization of the program. Most of them left within a week. New recruits were trucked directly to isolated camps before they had a chance to talk to anyone, but they still left.

Increasingly desperate as the rice rotted and sprouted on the ground, the bank raised the rate in stages and finally paid 1.25 per kilo, nearly half of the value of the grain, but to no avail. The season's losses were distributed among the insurance company, the bank, the commission, and SESA, and certainly ran well above the 11 million pesos reflected in the bank's figures. General confidence in the program was badly shaken. Reports went back to Ojitlan of a big failure in Uxpanapa, strengthening the hand of the opposition faction. The director of the agricultural program resigned.

1977 - The Problems Persist

The next year, the commission reduced the area to be planted by 50 percent, hoping to be able to organize the same basic system more efficiently on a smaller scale. As every slash-and-burn farmer knows very well, the longer tropical forest soils are cultivated, the more weeds, pests, and diseases threaten the crops. Costs inevitably go up unless a stable and efficient agricultural system can be found. The devaluation of the peso in 1976 increased the costs of inputs, but the prices for basic grains were kept low to control the inflationary spiral in the cities. The 1977 rice program in Uxpanapa was affected by all of the elements in this squeeze.

Once again, the land preparation was hurried. In theory, this operation should be performed at the beginning of the dry season, to expose the soil to the sterilizing effects of the sun (Mexico, 13). Time did not permit this, and the fields were not sprayed for soil pests, an increasingly serious problem. The disks did not break up the surface well enough to kill the weeds and the volunteers from the previous years' crops. Johnson grass, an especially tough weed resistant to most herbicides, was accidentally introduced with the rice seed. Blast attacked again and caused significant damage, although a better mixture of fungicides was applied. Other diseases, insects, and pests including birds, were even more concentrated than the year before. There was a short drought in August that seriously affected the plants. Again, there were not enough functioning combines to harvest the grain that did mature. The average yield was half what the same variety, Bluebonnet-50, had produced in 1976.

1978 - With or Without the Ejidatarios

Neither the ejidatarios, the insurance company, nor the Bank were willing to try again in 1978. The Papaloapan Commission agreed that large scale rice production would have to wait until more resistant varieties were found and a more easily controlled system was developed. Credit was approved for less than 400 hectares of rice planted by hand on land prepared by machine. Although the fields were to be located in compact blocks, each ejidatario was to be responsible for one or two hectares, or whatever he thought that he could cultivate and harvest by himself. Bluebonnet-50 was to be planted at the much lower density of 60 kilos/hectare in rows, both to avoid Blast by decreasing the humidity of the microclimate, and to facilitate harvesting if the crop lodged. It was hoped that a modest program of this kind would give both the ejidatarios and the technical personnel a chance to learn and develop the most appropriate method for growing rice in the region.

The large public expenditures have made Uxpanapa very susceptible to pressures from other agencies of the government. A serious drought in Sinaloa and Oaxaca in 1978, which brought the irrigation reservoirs down to dangerously low levels, led to the acceleration of the policy of moving rice into rain-fed areas. The rain-fed agriculture district in Coatzacoalcos allotted Uxpanapa 4,000 hectares, more than the area planted in the disastrous 1976 season. The commission realized that it could not force the ejidatarios to take on that much credit when they all understood the risks so well. It decided to plant rice itself on ejidal land with hired labor. Any profits would be distributed equally among the members; any losses would be absorbed by the commission. SESA gave assurances that 150 combines would be available for the harvest.

As of this writing, I do not know how the season went. The ejidatarios have not worked well within the system that has been introduced. The Idle Lands Law allows the government to cultivate surplus area in ejidos. There is a danger that the commission's time, energy, and resources will be concentrated on its own operations, that this same formula will be repeated in subsequent seasons, and that the collective ejido will no longer be the basic unit of production in Uxpanapa.

MAIZE

Both the Chinantecs and the spontaneous settlers are fundamentally maize farmers. This grain is the basic element in their diet. Their traditional systems are designed to provide the best possible growing conditions for the crop in a variable and difficult environment. They are capable of producing modest yields on small plots without any capital or technical assistance at all. Mexico is experiencing shortages of domestically-produced maize, because it yields poorly and receives a lower price than alternative crops in most parts of the country. Less than ten percent of the eight million hectares planted in recent years are cultivated using modern technology. The government has mounted a major campaign to introduce hybrids and modern cultural practices into regions where the annual rainfall is greater than 700 millimeters, particularly in the Southeast. This policy is based on the assumption that it should be possible to triple national production simply by extending improved technologies to these areas (Wellhausen, 1976, p. 144). Whether this goal is realistic or not, it has put the Papaloapan Commission under a great deal of pressure to grow maize in Uxpanapa.

It is not an area well suited to maize production. The very high rainfall in the summer season, between 2,300 and 2,800 millimeters between June and December, affects development, favors the spread of pests and diseases, and damages the grain in the ear. The spontaneous colonists have had better luck in the winter season. They select plots where the soils have a high water-holding capacity, and plant in December before the end of the heavy rains. The plants have time to establish the deep rooting systems that enable them to survive the droughts of March, April, and May. Working on a much larger scale, the commission has not been able to choose appropriate soils or get the planting done in time to avoid drought stress. In the first four years, yields in both seasons in the fields grown with credit have averaged between 200 and 800 kilograms per hectare, less than half of what the traditional farmers are able to produce. These figures include areas that were never harvested and reflect organizational as well as biological limitations.

Hybrids and Native Varieties

Over the past 35 years, both foreign and Mexican plant breeders have been attempting to improve the maize plant, with only limited success. Over thousands of years of continuous production, the campesinos have selected open-pollinated native varieties that are well adapted to the very specific ecological conditions and the pests and diseases of each small region. Hybrids, which revolutionized production in the United States, are produced from a deliberately reduced genetic base to perform extremely well under a narrow set of conditions. There is no natural corn belt in Mexico; the country, particularly the tropics, is divided into many distinct ecological zones. This does not permit the widespread adaption of a single hybrid strain. Hybrids cost more to produce than native varieties, both because new seed must be purchased every year and because high planting densities, fertilizer, and chemical pest control are necessary to create the conditions favorable for high yields. Improved open-pollinated varities take many years to develop and face similar problems.

In the first seasons in Uxpanapa, the bank and the insurance company mandated the use of *H-503* and *H-507*, which were developed in the early 1950s and are the most commonly used hybrids in the tropics. The first is a dwarf that performed badly and was dropped after one season. *H-507* has suffered from excess moisture, drought, helminthisporium disease, ear worms, and other pests. Its major drawback is that the leaves of the ear do not completely close at the point, allowing moisture and insect larvae to enter. Starting in 1976, the commission decided to buy native varieties of the *Tuxpena* type from the spontaneous settlers in the hope that they would do better adapted to local conditions. The seed was selected for uniformity and treated by the national seed company. The plants are hardy and completely enclose their ears with tightly wrapped leaves. Nevertheless, a large-scale operation does not permit the careful plot selection and husbandry practices the traditional campesinos have developed. Although the fields have been planted at high densities, fertilized, and sprayed, the yields have been too low to recover the costs of production.

The Labor Shortage

The greatest single problem limiting maize production during the first years was a severe shortage of labor. Maize was grown with a semimechanized system requiring between 30 and 50 man-days of labor irregularly distributed through the year - between two and three times that required for the rice (fig. 8.1). The fields were prepared and furrowed with tractors; field operations from planting through harvesting were performed by hand using methods borrowed from the traditional technology of the tropics. Most of the area was planted by espeque. A faster method was to drop the seeds directly in a furrow and cover it with one's foot, but this only worked if the campesinos walked immediately behind the tractor and planted while the furrow was still moist. Using the traditional system, a skilled man finds moisture at a slightly different depth with each push of his stick, which also lightly compacts the soil and provides a good microenvironment for germination. The stands planted in furrows were very uneven. Fertilizer was applied in a separate operation.

The resident population was not large enough to take care of the fields that had been cleared and prepared. Groups of the Chinantecs were brought down from Ojitlan to plant and fertilize. They worked as rapidly as possible to maximize their cash advances, which were paid on a per-hectare basis, and then went home to tend their old fields. The commission and the bank were left with insufficient manpower to accomplish vital, but intermittent tasks, such as replanting poorly developed patches, weeding, and applying chemicals.

Furthermore, as figure 8.1 shows, the maize harvest periods conflicted with other, more pressing activities; the Spring planting in May and June, the rice harvest in October and November, and the Winter planting in December and January. When the campesinos finally could get into the fields, they frequently found that the crop had deteriorated significantly. If it became clear that the value of production would not pay the accumulated costs, the last, preharvest inspection by the insurance company became especially important.

There is a key yield below which the value of the grain will not pay for the cost of harvesting and transportation. If the agent estimated a yield slightly above this, the ejidatarios had to turn their entire production over to the bank to meet their obligations. If he set a slightly lower figure and declared a total loss, they could write their indebtedness off to the insurance, sell some of the grain to private merchants, and keep what they needed for food. It is not surprising that many ejidatarios went out at night to pick or even destroy some of the maize before the inspector arrived. This was illegal and constituted "stealing" from their own fields. Serious conflicts of this kind obliged the insurance company to cancel the policies of some ejidos. The campesinos refused to harvest simply to turn the product over to the bank, grain rotted in the fields, debts went unpaid, and further credit was denied.

The technical staff, which was working nonstop to keep their system functioning, blamed these labor problems on the irresponsibility and laziness of the campesinos. The Chinantecs and other settlers, who of course take much better care of their own crops, complained about their small incomes and the low yields of the government's expensive

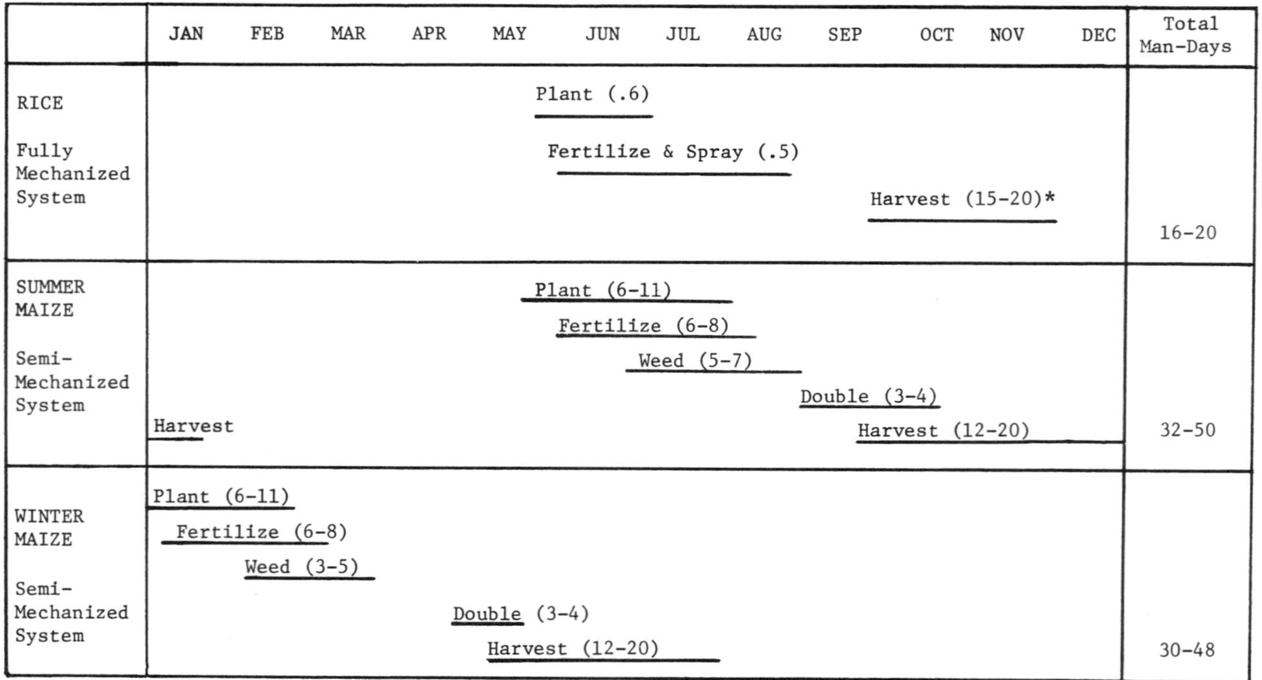

*Bagging and Loading Only. Figures in parentheses are man-days per hectare.

Fig. 8.1. Uxpanapa: Seasonal distribution of labor requirements for annual crops.

Source: Estimates of the Agricultural Program Office of the Papaloapan Commission and Interviews.

system. At the same time, the people needed maize to eat as more and more of them moved permanently into Uxpanapa. They insisted on the right to plant small plots for their own use without credit, which of course diverted even more labor from the commercial operations.

Increased Mechanization

The Papaloapan Commission has tried to solve these problems by increasing the degree of mechanization. Mechanical harvesting would prevent the long delays after the grain matures, which would certainly increase yields if it could be managed effectively. Combines require perfectly straight rows, and the campesinos have been unwilling to plant slowly and carefully with string lines at the rate they have been paid per hectare.

The first experimental plantings with machines were unsatisfactory, because the uneven terrain and the trash left over from the clearing made them skip and place the seed too close to the surface. As the residue decomposes, mechanical planters may be able to work effectively on the more level fields. This would allow the winter crop to be sown before the end of the rainy season, assuming that the land could be prepared on time and that the machines could work in a season when the soil is completely saturated. In 1978, the Union of Ejidos bought 27 planters, 30 maize heads for SESA's combines, and three large grain dryers to reduce post-harvest losses. CONASUPO, the state buying agency, planned to open reception centers in the zone. Nevertheless, it was still uncertain whether maize would ever be economically viable in the region on a commercial scale.

9 Perennial Crops and Livestock

The agricultural program in Uxpanapa has emphasized annual crops in the first years. Nevertheless, all of the technical studies the Papaloapan Commission has made have recommended that the forest be replaced with tree crops and pastures. They permit the soil to be kept covered at all times, and their root systems are a protection against erosion. The collective ejidos, which should combine the social advantages of smallholder production with the economies of scale of large plantations, were considered particularly appropriate for enterprises of this kind. Furthermore, the necessary processing, storage and marketing facilities could be centralized under the auspices of the Union of Ejidos.

It was important to choose crops and systems of production with a high probability of commercial success, because they require large investments that do not begin to pay off for several years. Detailed knowledge about the common tropical plantation crops has been accumulated in many parts of the world. In addition, the economic potential of many other species that have long been ignored has received increasing attention in recent years (National Academy of Sciences, 1975; Romanini, 1976). In tropical Mexico, the best choice in the short run is the production of perennial crops, which can be substituted for foods or industrial raw materials that are currently either imported or produced at relatively high cost under irrigation. Natural rubber fits these criteria well, and a program to plant 5,000 hectares was initiated in Uxpanapa in 1975. Other possibilities, including sugar cane and African oil palm, are being investigated.

Ranching is the most widespread enterprise in the Southeast, but the average efficiency of the operations is very low. Demand for both meat and milk is expanding rapidly in Mexico, especially in the mushrooming petrochemical centers along the Gulf coast. Intensive livestock operations are being developed in Uxpanapa as rapidly as the institutional structure will allow.

RUBBER

Rubber is produced by a variety of trees and shrubs. It was used for balls and other articles by the native peoples of Mexico before the Spanish conquest. The Olmecs, the earliest important culture, are known by their Nahuatl name, which means "the people of the rubber country." Latex was extracted from the Castilla elastica tree, and formed an important part of the tribute paid to the Aztec empire by its subjects in the tropical lowlands (Rodriguez, Penalosa, 1975, p. 3).

Almost all of the natural rubber produced in the world today is tapped from one kind of tree, Hevea Brasiliensis, which is found wild in the Amazon basin. Raw rubber becomes sticky when it is hot, so that it was little more than a curiosity until George Firestone patented the vulcanization process in 1837. The introduction of rubberized clothing at the time of the American Civil War increased demand, but it was the rubber tire that made it an important item in world trade. Enormous fortunes were made in Brazil at the end of the nineteenth century from the collection of latex from wild trees.

Almost the entire stock of the modern plantations is descended from seeds Sir Henry Wickham collected in Brazil and sent to the Royal Botanical Gardens in Kew, England in 1876. H. N. Ridley, the director of the botanical garden in Singapore, almost single-handedly developed the basis of the modern cultural system, by grafting high-yielding clones onto vigorous root stock and by devising a tapping method that increases the flow rather than damaging the trees. Plantations spread very rapidly to meet growing industrial demand. Southeast Asia now produces nearly 90 percent of the world's natural rubber (Polhamus, 1962, Chapter III; Purseglove, 1975, pp. 148-51).

The establishment of plantations in tropical America has been limited by a fungus disease, South American leaf blight (Dothidella ulei, or Microcyclus ulei) which has never spread to Asia. The problem illustrates the danger of growing a wild plant in monoculture. Only 12 to 14 Hevea trees are found per hectare in their native habitat. This very low density prevents the spread of spores and the disease is rare. When the somewhat more susceptible high-yielding varieties are brought together in one area, the pathogens are concentrated and the problem becomes very serious. An 8,000 hectare plantation set up in Brazil by the Ford Motor Company, Fordlandia, was destroyed in the 1930s; and early attempts to introduce intensive cultivation into Mexico failed (Rodriguez, Penalosa, 1975, p. 4; Romero, 1898; Thurston, 1971, pp. 181-6).

History in Mexico

During World War II, the Allies were cut off from the major producing centers by the Japanese. The United States Department of Agriculture sponsored an intensive program to establish productive plantations in

Mexico and Central America. Varieties partially resistant to leaf blight were found, and valuable information on cultural methods was collected and tested experimentally. The same strategic shortage stimulated the development of technology for the production of synthetic rubber from petroleum. After the war, Mexico could not compete with the lower costs of production of both the Asian plantations and the international oil industry. A modest development program under the auspices of the Papaloapan Commission was promoted without much enthusiasm in the 1950s.

By 1961 national rubber consumption was increasing rapidly. To reduce the growing import bill, the government set up a natural rubber program financed by a tax on imports. It was administered first by INIA and later by the Mexican Coffee Institute, as a complement to its other activities in the same areas. New plantations were set up and credit and technical assistance were provided to growers during the six years of the program.

Support was withdrawn at least in part because the small group of processing industries, dominated by subsidiaries of the big American tire companies, insisted on importing their raw material. Ninety-five percent of the natural rubber imported into Mexico comes via the United States, which of course is not a producer. National production peaked at about 5,000 tons, less than 7 percent of demand, in 1974. Two synthetic rubber plants were opened in 1967, but natural and synthetic rubber are not perfect substitutes for all uses. They are combined in the manufacture of tires, which account for 65 percent of demand. Natural rubber imports are expected to rise to over 60,000 tons by 1980 (Rodriguez, Penalosa, 1975, pp. 49-56; Shepherd, 1975).

After credit for new plantings was discontinued in 1967, the producers received erratic prices and very limited technical assistance. The experiment station at El Palmar, which has worked on rubber since the 1940s, has been funded sporadically and has not kept up with improved varieties and cultural methods. Only small farmers and ejidatarios with less than five hectares of trees and very limited alternate opportunities have kept up with the management practices necessary to maintain adequate yields. The national average of 625-700 kilos of dry rubber per hectare is very low by world standards. Yields in Malaysia, the world's largest producer, range between 1,000 and 1,500 kilos on small holdings to over 2,500 kilos on large plantations. The costs of production in Mexico are 30 percent higher than the Asian average (Purseglove, 1975b, p. 170; Shepherd, 1975).

Fresh latex should be transported and processed as quickly as possible under perfectly sterile conditions, in much the same way as fresh milk. There are seven small processing plants in the Southeast with a combined capacity of 755 metric tons per month, or about 7,000 tons over the nine to ten month tapping season (Plancarte, Alcarez, 1977). Fluctuating prices, and the logistical problems of collecting from scattered plantations, have prevented the organization of efficient processing systems. Using very simple equipment, the plants turn out smoked sheets, crepes, and granular rubber from latex coagulated and stored for as long as a week in the field. Different grades are mixed into a homogeneous product of inferior quality. Domestic rubber is used primarily for shoe soles and other simple items with low specifications (Rodriguez, Penalosa, 1975, p. 32).

The energy crisis of the early 1970s, which pushed up the price of both synthetic and natural rubber in the world market, has led to a new program for the development of plantations, spearheaded by the Papaloapan Commission in Uxpanapa. A special rubber authority, financed through the official bank, plans to establish nearly eighteen thousand hectares in the Southeast, more than tripling the area in production, and to create new processing plants and other services. Most of the world's rubber is grown within 10° of the equator; the natural conditions in Mexico are not ideal. The trees mature more slowly and yield less than in Africa and Asia. Higher labor and other costs will almost certainly prevent Mexico from competing in the world market. Nevertheless, if the new plantations can produce a ton of dry latex per hectare, it is believed that they could supply as much as a third of domestic demand (Tamayo, 1976b).

The Program in Uxpanapa

The rubber program in Uxpanapa has been organized rapidly under a great deal of pressure. The Papaloapan Commission, in cooperation with the experiment station at El Palmar, had produced over four million young trees ready to be transplanted into permanent plantations in the ejidos by 1978. Seeds were germinated in shaded beds, and the seedlings were established in densely planted nurseries. After a year, they were bud-grafted with high-yielding clones resistant to leaf blight. Budstock loses it viability after a few days, and was transported from El Palmar, over 400 kilometers from Uxpanapa. Grafting is impossible in the rain, and it was difficult to insure that enough material and trained personnel were available when conditions were favorable. The proportion of successful grafts was disappointing, and the nurseries were damaged by leaf blight, insects, moles and other pests.

Seedlings should be established in plantations within a year of the time they are grafted. This pressure built up rapidly, before appropriate sites had been identified and prepared, and before the ejidatarios had been organized into production teams. A tap root between three and ten meters long and an extensive lateral system allow the *Hevea* tree to tolerate infertile soils, if they are deep and well-drained (Moraes, 1977, pp. 317-318). These conditions were met in the relatively flat areas in Uxpanapa that, however, had already been cleared for annual field crops. The soils on the steeper slopes, designated for perennial crops in the original plan, were too shallow. As more detailed studies were made, it was difficult to find suitable areas large enough for efficient plantations - between one and three hundred hectares.

One of the greatest theoretical advantages of rubber within the context of the Uxpanapa resettlement program is that it can provide a steady source of year-round employment. The care of the trees from the time the stand is planted until it goes into production requires an average of 35 man-days per hectare per year. Two people working together can tap about 500 trees per morning, nine or ten months a year. The complete care of a mature stand requires between 165-200 man-days per hectare per year. An average family can provide 1.7 workers or 530 man-equivalents; enough to care for two or three hectares with very little surplus time for other activities. If all the work is to be done by the ejidatarios themselves, nearly 2,000 families will be required for the planned 5,000-hectare program.

Plantation agriculture will involve a radical change in the lives of the Chinantecs, who have always been independent farmers on a small scale. During the first years of the Uxpanapa project, the technical staff was extremely busy with the annual crops, and were unable to organize or train the ejidatarios in preparation for the rubber program. In 1976, not one of the ejidal presidents interviewed for this study knew what the program would involve or expressed any interest in growing rubber. The previous program in the 1960s was centered in the Tuxtepec area, and many of the people had heard stories about low prices and erratic official support.

As the pressure to establish the permanent plantations built up over the next few years, meetings were held in the ejidos where suitable, compact areas had been identified. The representatives of the commission, the bank, and the new rubber authority did little more than promise millions of pesos in credit advances over the next decade, followed by guaranteed high incomes. The Chinantecs were moving into Uxpanapa very slowly, and only a minority of the members of the new ejidos were in residence. They had not worked out the details of their organization; who would be directly involved in the rubber program, how the labor would be organized and paid, and how the plantations would be divided up. The ejidatarios were disillusioned by the poor yields and low incomes of the rice and maize programs, and accepted long-term credit for rubber if they could meet the requirements. Several spontaneously settled ejidos where large areas had already been cleared by hand, and which were well organized under cohesive leadership, contracted for credit. By 1978, over 1000 hectares had been transplanted. The new plantations will take about eight years to come into production, so that there will be time to work out the details. Nevertheless, a new crop with complex management requirements is being introduced from the top down, according to inadequately formulated technical criteria, without the effective participation of the ejidatarios.

LIVESTOCK

Livestock production will become increasingly important in Uxpanapa over the next few years. The Papaloapan Commission is using the project as an opportunity to introduce efficient, modern operations, which are rare in the Southeast. Since Colonial times, more land in the lowland plains has been dedicated to pasture than to any other use (Poleman, 1964, pp. 72-75). About a third of the approximately 28 million head in Mexico are raised on 192 million hectares of pasture in the tropics (Ramirez, 1974, pp. 28-29). Most of the ranches are extensively and inefficiently managed, although there are some intensive operations that are integrated with fattening and marketing operations in other parts of the country. Cattlemen tend to concentrate on either rearing or finishing operations, depending on the natural conditions and the level of capital investment. Milk is a sideline; there are very few specialized dairies. The principal market is in Mexico City, although the prosperous, rapidly growing petrochemical centers along the Gulf coast have significant potential.

The most common animal in the tropics is a cross between a native criollo type, descendant of the cattle introduced from Europe in the sixteenth century, and Cebu cows from India. In recent years, temperate breeds, particularly Brown Swiss, have been introduced to upgrade the stock and improve milk production. The many complex factors involved in the efficient management of cattle in the rainy tropics are beyond the scope of this study (McDowell, 1972; Williamson, Payne, 1965). Constantly high temperatures and humidity put the animals under stress and favor the development of diseases and both internal and external parasites. The grass is green all year around, but its nutritive value declines significantly in the dry season. Native breeds are well adapted to these conditions, but their growth, reproduction, and milk production rates are low. Artificial breeding, improved pastures, careful disease and parasite control, and the other practices that are necessary if better animals are to achieve their genetic potential, are being introduced very slowly. The average stocking rate in the Southeast is under one animal unit per hectare, and the average returns are low (Mexico, 20; Osorio, n.d.).

About 75 percent of the cattle are privately owned, many of them by absentee landlords who live in the towns or in other parts of Mexico. Especially along the forest frontier, the ranches tend to be large. Most ejidatarios are primarily dedicated to crop production. The division of the land into small plots, the need to produce much of their own food, and very limited capital resources do not permit livestock operations on a large scale. Families keep draught animals and domestic species such as pigs and turkeys, but only rich ejidatarios raise cattle, both as a status symbol and as a way to invest surplus capital. The few ejidos almost entirely dedicated to livestock are either located on land which has become too exhausted for crop production, or are contracted by local ranchers to raise their animals (Mexico, 31). Few government development or credit programs have been able to overcome the complex economic and social problems that hamstring the ejidal sector, and lay a solid groundwork for efficient, profitable enterprises.

Cattle Operations in Uxpanapa

In cooperation with FIRA, a specialized agency of the Bank of Mexico that finances long-term investments in agriculture, the commission has set up a few intensive operations in Uxpanapa based on improved animals and pastures. The basic plan for the first of these was established in the ejido "Almanza" in 1977. The capital investment for the first ten years of operation is four million pesos, financed almost entirely with credit. The enterprise is designed as a double-purpose system; live beef animals will be sold to the Mexico City market, and milk will be produced for local consumption. The animals are improved crosses between Cebu, Brown Swiss, and Holstein cattle.

Except for mineral supplements, particularly phosphorus, grazing will be the only source of feed throughout the year. African Star Grass (Cynadon plectostachyum) has been selected as the principal forage grass for the region. Evidence from other parts of Mexico and Puerto Rico (Vicente-Chandler, 1974) suggests that this is a good choice. It is established vegetatively and rapidly covers the ground by extending its stolons. It does not flower and is less subject to rapid changes in nutritive composition than many taller species. It responds well to nitrogen fertilizer. It grows back readily after grazing and trampling, need not be mown or burned over, and is relatively resistant to both drought stress and insect attack. A grass-legume mixture will be established in a smaller area to provide additional protein to young stock and lactating cows. The pastures will be rotated, fertilized, and weeded to prevent the invasion of undesirable species and woody shrubs.

When the herd stabilizes after the sixth year of operation, the ranch will support over a thousand head of cattle at twice the average stocking rate for the Southeast. A permanent work force of four men and a part-time veterinarian will be expected to manage a profitable operation.

As was noted in connection with the rubber program, all of the agricultural enterprises in Uxpanapa have been competing for the best land. According to the original plan, cattle were to graze slopes greater than ten percent, which would be cleared by hand. The technical representatives of FIRA have not been very impressed with the results of the rice and maize programs, and have been unwilling to approve large investments for livestock operations unless they are located on good soils with easy access to water. The commission has been unwilling to write off its 1,000-dollar-per-hectare investment in mechanized clearing, although the "Almanza" ranch is located on land which they cleared. This and other institutional conflicts have slowed the establishment of cattle operations in other ejidos.

Once again, the government visualizes the introduction of a complex, sophisticated, and untested technology into Uxpanapa. The projected performance levels are relatively high, and there may be problems organizing the market, especially for milk. If the credit is to be paid back and the enterprise profitable, excellent management will be needed. Many of the wealthier Chinantecs did keep cattle in the

Cerro de Oro lakebed, but none of them have any experience in modern husbandry techniques. The professional technicians and veterinarians of the government agencies will be able to manage one or two pilot operations themselves, but they will not be able to keep track of all the details if many ranches of this kind are established in the zone.

Very little emphasis has been placed on an alternate policy - on the development of low-cost systems based on sound but less sophisticated practices. In 1978 there were already over 3,000 head of mixed cattle in the zone. Some of them had been brought in by the Chinantecs; the others belonged to the spontaneous settlers. The ejidatarios have been frustrated that neither credit nor their own funds, which are deposited in the care of the Union of Ejidos, have been available to expand and improve these operations.

They would prefer relatively simple systems based on Cebu-criollo crosses and hardy native grasses. The cattle would not grow as quickly as the improved breeds under ideal conditions, they would lose weight during the dry season, and they would give less milk. Nevertheless, careful management practices such as pasture rotations, supplemental feeding of agricultural wastes, and intensive training programs could make the operations profitable. Improved animals and forages could be introduced when the people were capable of managing them themselves (McDowell, 1972; Staver, 1977).

SUGAR CANE, OIL PALM, AND OTHER EXPERIMENTS

There is more experience with sugar cane production in the Southeast than with any other perennial crop. The Papaloapan Commission has set out plots to test the performance of a number of Mexican and Cuban varieties in Uxpanapa. As would be expected in a newly cleared area, experimental yields have been high: 105 tons of cane per hectare with a 19 percent sugar content. The crop is a very high consumer of nutrients. In other areas with similar soils and environmental conditions, such as the fields associated with the Lopez Mateos mill near Tuxtepec, yields have fallen off rapidly after a few years, and erosion has become a serious problem (Rodriguez, 1977a, p. 492). Detailed technical studies have not yet been made in Uxpanapa, but it has been estimated that between two and five thousand hectares would have to be planted to justify the construction of a small mill. The crop would require the best soils available.

When cane is grown on ejidal land in other areas, all of the mechanical field operations are performed directly by the mill on a contract basis. The ejidatarios, as owners of the enterprise, are unwilling to perform the back-breaking labor of cutting cane. Outside workers are brought in, and the ejidatarios live off the rent without taking any risks or doing much of any work at all (Mexico, 3).

African oil palm (Elaesis guineesis) is a native of the rainy tropics and is one of the most productive sources of high-quality oil in the world. Mexico is importing increasing quantities of both edible and industrial oils. Preliminary studies have concluded that it could be a profitable enterprise in Uxpanapa, although a shortage of direct sun may limit yields (Palma Gomez, 1977). There is almost no experience in its culture in Mexico; it will be necessary to train technicians and run field trials before capital is committed on a large scale. Oil palm is a promising long-term prospect, if the organizational problems of operating a complex plantation system can be worked out in the rubber program.

An experimental program is planned to test other possibilities, including the following:

Annuals	Perennials	Spices	Pastures
Maize	Sugar Cane	Cloves	African Star Grass
Rice	African Oil Palm	Cinnamon	Bermuda Grass
Beans	Bananas	Pepper	Santo Domingo Grass
Soy Beans	Avocados	Vanilla	Jaragua
Sorghum	Mangos		Privelegio
Chili	Zapote		Aleman
Pineapple	Mamay		Various legumes
Various vegetables	Coconuts		
	Coffee		
	Lemons		
	Oranges		
	Grapefruit		
	Cocoa		
	Cassava		

Four hundred hectares for a central experiment station have been cleared in the valley of the Uxpanapa river on some of the best soils in the area. Planning will be coordinated with INIA to avoid the repetition of experiments that have already been done elsewhere. The development of an ambitious program of this kind is slow, especially as the young and inexperienced staff does not have easy access to library or laboratory facilities. Nevertheless, Uxpanapa may become an important research center for tropical crop production.

The Intensive Garden Program

All of the commercial enterprises in Uxpanapa are operated on a large scale on a collective basis. The 1600 square meter plots behind each house in the new towns are the only privately-operated land in the system being introduced. An experimental model for the intensive home production of vegetables, pigs, chickens, and rabbits has been established to replace the back-yard gardens that played an important role in the family economy of the Chinantecs in Ojitlan. The commission would like to integrate both production and marketing through an efficient, centralized structure. There have been technical problems with the temperate vegetables that have been introduced, and the improved breeds of rabbits have not prospered. The ejidatarios have shown very little interest in the program so far.

10 Collective Organization and Modern Agriculture

The helicopter hovered low over the treetops, banked sharply, shot through a patch of mist, and crossed back over the edge of the jungle to the perfectly rectangular, cultivated fields. The contrast was extraordinary; the brilliant green of the young rice plants looked like a golf course improbably laid out in the wilderness. Each field was surrounded by the darker color of the ragged windbreaks, and was dotted with eight, evenly spaced, black piles of trash, some of which were still smoking from the recent burn. We passed over several areas not yet planted. In one of them, a bright orange tractor crept along at the head of a swath of newly disked, dark red earth. The agronomist, who was sitting in the front seat, shouted something into the ear of the pilot over the shrill whine of the jet engine and the roar of the blades. A few moments later, we set down in the middle of a field, creating a small hurricane which whipped at the young maize plants. We got out, ducking low, and walked over to a dirt road where a group of Chinantecs were resting beside a white dump truck, leaning on their long, hand-sharpened planting sticks. They were the group from Paso Limon, who, in 1976, were held up as a model of collective consciousness and hard work, and who symbolized the potential of Uxpanapa.

We shook hands with everybody, and were introduced to the leader of the group, a young man of about thirty. The agronomist asked him some questions for our benefit: "How many of you are working here in Uxpanapa?" The leader smiled and answered, "thirty-five." "How many hectares are you growing this season?" "Seven hundred: 610 of rice and 90 of maize." "How many hectares did you cultivate in Ojitlan?" "It's difficult to say, two or three hectares each, maybe 80 between us." The agronomist laughed and slapped him on the back. We joked and chatted for a few moments, walked back to the helicopter, and continued on our tour.

I visited these people many times in the next few months; nearly every day during the rice harvest. Paso Limon is one of the largest and most prosperous ejidos in the Cerro de Oro lakebed, and is located near the ancient Chinantec center of Tlacuache. It has had a primary school since the 1930s, and produces some of the highest yields in the region of rice and tobacco on an alluvial plain that is rapidly being eroded away by the Usila River. The resettlement program has been very controversial in this village, and the community is bitterly divided into factions.

In the first stage of the Uxpanapa project, the Papaloapan Commission cleared over 1,500 hectares of good soils at the eastern end of the zone. New ejidos had not yet been formally organized in this area, which was far from the new towns being built first and had never been settled by spontaneous colonists. The commission and the bank organized credit associations of people who would eventually move there to cultivate the crops. The group from Paso Limon were members of a minority in the village who had been converted to evangelical protestantism. They were strongly unified already, and were part of the pro-Uxpanapa faction. They built temporary dormitories for themselves on the banks of the Uxpanapa River, and moved in without their wives and children. They all looked forward to the day when they would confront their adversaries with big profits as proof that the Uxpanapa project would benefit all of them if they would work together with the government. They signed credit contracts for over three and a half million pesos for the first season.

Although more rice was salvaged from their fields than from any other part of the zone, the season was a tremendous disappointment. Pest and disease control alone cost over a million pesos. The rice lodged, and neither the combines nor the hired workers who were brought in were capable of harvesting enough to pay the costs. Maize yields were also very low. Nearly a third of the group went home at the end of the season.

When I visited them the next year, their mood was subdued. The bank has given them credit for a new truck of their own, but they were losing faith in the project. The replacements who had been convinced to come from Paso Limon refused to work for the cash advances offered in the credit contracts, and had planted small plots of maize without credit to subsist. The leader took me out to the fields, which were heavily infested with weeds. We had a long conversation about the collective system, in which he expressed very serious doubts about the possibility of getting the people to work together under these conditions. His own personal position in the community was being undermined, and he felt that he was losing control of the group.

By the end of 1977, the spirit of the group was entirely broken. The rice crop was very small once again. Opposition leaders from Ojitlan came and convinced the people not to harvest the maize, simply to turn it over to the bank. The insurance company refused to honor the loss, and the bank was unable to offer more credit until the debt was repaid. By that time, the commission had started to work on the new town in which the settlers from Paso Limon will live. Those members of the original group who did not return home lived on the subsidies they received for building their houses. There was nothing to do except wait until the project was completed, and see what the future would bring.

In 1978 on my last, brief visit to Uxpanapa, I was invited to attend a meeting held in the dining hall of one of the construction camps. All of the tables were pushed together into one long row. The resident director of the project and representatives of the agricultural program office, the bank, and the Agrarian Reform Ministry came in together and sat at the head. This ejido was composed both of spontaneous colonists who were living in their original village, and a number of Chinantecs who were moving into one of the new towns. There was a delay while a truck was sent to fetch some of the Chinantecs who had missed their ride.

When about forty people were settled, the director stood up. He said that he knew that there were problems and misunderstandings in the ejido. Several groups had been to see him in his office, and he had called this meeting so that everybody could participate in finding solutions. Mexico was having severe economic problems, the federal budget was restricted, but Uxpanapa was high on the list of priorities of the president of the Republic. Just that day, he had received a telegram from Ing. Jorge Tamayo, executive director of the Papaloapan Commission. It authorized the construction of new houses for the ejidatarios, not all at once, but at a steady rate. He had a copy, which he would like to take the opportunity to deliver personally to the president.

The president stood up. "How many people think that I should sign this receipt?" Only about six people raised their hands, and the ejidatarios turned to each other and began to discuss the problem. "Sign it!" shouted a few. "No," said others, "we don't want to move to the new town, we want to stay where we are!" The president turned to the director, and said that the people he represented did not want him to receive the document.

The officials were amazed by this rebuff, and consulted among themselves. After a few moments, the representative of the Agrarian Reform Ministry stood up. "I think that most of you know me, although I have been here in Uxpanapa for only a short time. I would like to point a few things out to you. The Papaloapan Commission cannot force you to do anything. It is here to develop the potential of resources which are very important for Mexico. In return, it is offering you a chance to participate, and is providing you with many services. We cannot make you live where you do not want to live, but I am sure that at least some of you would like to move to the new town. We are in a position to offer 20,000 pesos per household, electricity, and many other things. By accepting this document, you are being given an opportunity. It is like a registered letter, which you must sign for at the door before you can read it. The receipt is simply a record that we have done our duty and delivered it to you. You are not committing yourselves to anything."

The ejidatarios discussed the issue heatedly for about fifteen minutes and asked many questions, some of them in a hostile tone. At last, they decided to get on to more important things. The president signed the receipt and put the envelope in his pocket without opening it.

The director stood up again. He said that at this time of year, in the early Spring, the most pressing business was to plan the next cropping cycle. The Papaloapan Commission had been entrusted with the development of Uxpanapa, and millions of pesos had been spent. Everyone who had been given the great opportunity to live and work there, officials and ejidatarios alike, had an obligation to make the area produce for the people of Mexico as a whole. They had come this evening to offer new lines of credit for rice and maize.

The ejidatarios started speaking all at once, both in Spanish and in Chinantec. After a few moments, the president calmed them down. He turned to the head of the table and said that his people did not want to work with the bank any more, and did not like the collective system. They had never made any money, there had been nothing but losses, and many of the companeros were already preparing small plots to cultivate in their own way. They only asked to be left alone.

The director was expecting this, and spoke at some length. As they all knew, and as was only to be expected in a new project, there had been a variety of problems in the first years. Little by little, solutions were being found. For example, the quality of the land preparation had been poor, because SESA had used the wrong kind of disk. The new machinery station, under the control of the ejidatarios through the Union of Ejidos, would do better. In past years, they had relied too heavily on the airplanes to fumigate, and the costs were going up. This year, they would obtain backpack sprayers to give work to the ejidatarios. Poor communications had resulted in many problems with the insurance company; an arbitration office was being set up in Uxpanapa to find solutions on a case-by-case basis.

In this season, they were asking only those ejidatarios who wished to participate to form production teams, which would receive credit and technical assistance. Each member would decide how much rice or maize he felt that he would be able to take care of and harvest. This way, they would not be dependent on the combines. Each group would plant in a compact block of land which would be prepared all at once by the new tractors. They would share responsibility for the credit, but were not being asked to take on more than they could handle.

This new idea provoked much discussion in the group. One man stood up, and said that he did not want to work with the Bank under any conditions. He had been working for the past month preparing a plot, where he intended to plant his own crops in his own way. The director asked him if he had cleared the land himself, or if he was using an area that had been cleared by machine. The man hesitated, fearing a trap. He said that he had worked very hard clearing thick, secondary growth, although the machines had passed that way over two years before.

The director consulted with the agronomist in charge of the agricultural program. He then said that the machine-cleared areas could be planted without credit, as long as all of the ejidatarios who wanted to work this way stayed within one compact area of their own choosing. The commission could not allow cleared land to go to waste, and would use the area the members were not going to cultivate this season itself. They would bring in tractors and plant crops to meet Uxpanapa's production goals, and could not maneuver around scattered small plots. The profits of this operation would be divided equally among all of the legal ejidatarios, whether they participated in the credit program or not. He hoped that they had seen the last of their major problems, but if there were some natural catastrophe, the commission would absorb the entire loss.

This idea provoked even more discussion. One elderly Chinantec tried to make himself heard for several minutes. He spoke Spanish very badly, and was anxious to make a point he could not articulate. With the help of several intermediaries, he finally made his question clear: did this mean that the Chinantecs who had not yet moved from Ojitlan would lose their land?

The director quieted the group. He thanked the old man for asking this very important question. Every family affected by the Cerro de Oro dam had been given rights to twenty hectares in Uxpanapa by the president, and they certainly would not lose them. On the contrary, he strongly urged them to move in immediately. The commission would borrow the portion of the ejido's land the members could not use on a season-by-season basis. Within a short time, the ejidatarios would need all of their resources for crops, livestock, and other enterprises.

For the next hour and a half, the director went around the room asking each ejidatario how much land he wanted to plant with credit, how much without, and in what areas. Seventeen members signed up to join production teams which would plant a total of 196 hectares; the others preferred to work independently. The representative of the Agrarian Reform Ministry promised that he would be back the next day with maps to locate the areas in detail, and to write up a lease agreement for the over 300 hectares of cleared land the ejido was not planning to use. The meeting then broke up.

In the push to get the Uxpanapa project moving, the introduction of modern agricultural technology has not meshed with either the resettlement program or the establishment of viable cooperative farms. The new system as a whole, and each of its parts, are set up as if smoothly functioning, profitable enterprises already existed. The Papaloapan Commission has not been able to deal effectively with two dynamic processes: the transformation of the rain forest into an agriculturally-productive area, and the transformation of traditional Indian campesinos into modern farmers. It has not broken either process down into stages of programmed or orderly sequence. Serious conflicts have developed between these fundamental goals.

Although one of the principal arguments in favor of rapid development was to provide the Chinantecs with an established structure to move into, they have been coming very slowly. Some groups have even moved back to Ojitlan after a few years of unfavorable experience. A great deal of time and energy have been invested in the organization of the collective ejidos. Nevertheless, by 1978 many of the ejidatarios were disillusioned with this institution, and were anxious to work individually. The managers of the project, frustrated by their inability to organize a disciplined labor force cooperatively, were beginning to cultivate the land directly with hired workers.

The collective ejidos organized by President Lazaro Cardenas in the 1930s were intended to form the basis for agrarian socialism. Under the pressure of a general strike by workers with a high level of political consciousness, modern, technically-sophisticated haciendas were expropriated and transformed into production cooperatives. The same institution is being introduced into a very different social context in Uxpanapa to achieve entirely different goals.

Earlier colonization projects in the Southeast failed to meet their economic or social objectives because the government was unable to coordinate the interests of the settlers with the provision of an adequate infrastructure and the establishment of viable agricultural enterprises. Some of them, of which Los Naranjos is a good example, fell into the trap of excessive paternalism. Roads, housing, and other infrastructure were built at a cost of approximately 2,000 dollars per family. The colonists were entirely dependent on the commission, but the groundwork for successful development was not laid and the project was abandoned. Others, such as the program that opened Uxpanapa up to spontaneous colonization in the late 1960s, provided so few services that the people were forced to subsist on the basis of very primitive slash-and-burn techniques. In Uxpanapa, the Papaloapan Commission has invested heavily in infrastructure, over 20,000 dollars per family, to promote rapid development. The collective ejidos are intended to provide a framework for democratic participation and group responsibility, thus avoiding the problems of paternalistic dependency. So far, they have not been able to perform this function.

The Chinantecs, natives of an isolated valley with an indigenous culture of their own, are not the most logical choice for this kind of experiment. Like many Indian peoples of the South, they do have a certain tradition of collective labor. Nevertheless, the basis of their economy is the family, and they certainly would have preferred the individual assignment of parcels. The new cooperatives are complex organizations defined in detail by law, and require highly cohesive groups if they are to function smoothly and equitably.

There has been much discussion in Mexico about how native peoples should be treated in development programs. Their most ardent defenders, particularly some anthropologists, believe that the Indians should be protected from the corrupting influences of the capitalist market economy and the national culture. Many technicians, on the other hand, see the aspects of the indigenous cultures that have survived 450 years of domination by the mestizos as obstacles to be overcome in a necessary process of modernization. An anthropologist who has been very influential in the formation of national Indian policy commented on the Cerro de Oro project as follows (Aguirre Beltran, 1976, pp. 226-7):

> The proponents of total assimilation, by ignoring the traditional languages and cultures, will incorporate alienated, inauthentic beings--mere rags and tatters of humanity--into the general society. On the other hand, the extreme relativists would try to preserve these people as exotic museum pieces, which would lead to a policy of reservations . . . . We should not try to preserve the vestiges of colonialism which have contributed to a consciousness of inferiority and subordination. We must, however, conserve the native languages which are the symbolic systems by which the Indians express the values which give meaning to their lives.

The employees of the resettlement committee in Ojitlan made a tremendous effort to respect the traditional culture of the Chinantecs as the first stages of the program were organized. The emphasis in Uxpanapa, however, is on agricultural development. The project has been staffed almost exclusively by technicians from the north of Mexico. Very few of them had previous experience working with Indian people - most of them have never been in the Southeast before - and none of them speak the Chinantec language. The Agrarian Reform Ministry was not represented in the zone between 1975 and 1978. The agricultural program office of the commission and the bank were responsible for organizing the ejidos to meet the minimum requirements of the credit system. This process involved a great many meetings in each of the ejidos and endless paperwork, but both the directors and the small technical staff were under a great deal of pressure to produce rice and maize and to get the rubber and livestock programs organized. There was not enough time to work through the detailed questions and problems of each group. A few Chinantecs have learned to drive tractors and work in the rubber nurseries, but there have been no formal training programs in the ejidos.

The settlers were promised that they would be able to overcome their traditional poverty and become prosperous farmers in Uxpanapa with credit and mechanized technology. The low yields and heavy losses of the first years were demoralizing for everybody. It was very easy for the technical staff, which had been given the impossible job of managing the entire program, to fall back on a common racial stereotype and blame the problems on the laziness and selfish individualism of the Indians.

Under the very best of conditions, it would not be easy to transform the Chinantecs into disciplined members of complex organizations. Even if we put the cultural question entirely aside, the Uxpanapa project has not yet met the needs of the ejidatarios and has not provided an attractive alternative to their traditional agricultural economy. Credit has only been available within the context of the integrated program, over which they have had no control. The land in each ejido has been used according to the priorities of the government for the zone as a whole, not according to the wishes and needs of the ejidatarios. The choice of soils, varieties, planting dates, and many other technical decisions have been made contrary to the detailed empirical experience of the people, who have been growing the same crops in the tropics for years. The quality and timing of the mechanized field operations have been very poor. There have been many conflicts with the bank, and especially with the insurance company. There have been severe shortages of labor in certain months, and yet not enough work has been available through the year to provide an adequate income. The cash advances that have been paid on a daily basis have been very low, and in some cases have been unfairly distributed within the ejidos. The harvests must be marketed through the bank, often at a lower price than they would bring from private buyers. The costs of production have been very high, the yields have been low, and there have been no profits.

The paternalistic aspects of this system are not new to the Chinantecs; private credit in Ojitlan was also provided as a series of advances against the harvest, and their previous experience with official programs was unfavorable. Nevertheless, in Uxpanapa, the ejidatarios do not have the diversified crops and supportive social structure of their homeland. They have been living fairly well in the first years on their indemnification payments, their income from their plots in Ojitlan, and relatively highly paid work as peons for the commission. Although their real incomes are certainly higher than ever before, they are entirely dependent on cash for the first time in their lives. The agricultural program organized through the collective ejidos has not provided them with the big profits they were promised, and they feel that they have lost control over their lives.

The Chinantecs are disoriented, frustrated, and unhappy with the project in many ways, but they have no direct means of influencing government policy. They are divided politically into two factions. The only form of direct action the Mexican political system normally allows to the campesinos is the right to petition. At the time of the conflict over the Cerro de Oro dam, the affected population was promised a wide range of services by presidential decree (appendix). The government always promises more than it can deliver, and distributes its resources according to the economic importance of the various regions and the pressure the campesino organizations are able to exert. Many of the Chinantecs feel that the only effective form of leverage they have is to refuse to move to Uxpanapa until the infrastructure is entirely completed, and until the program is better organized. This position, which was expressed in many interviews, is not supported by the CCI (The Union of Independent Campesinos), the official representatives of the majority. The national leadership of this organization is strongly committed to the collective ejidos and has worked closely with the Papaloapan Commission. The CNC (The National Confederation of Campesinos) is the largest campesino union in the country and is a constituent element of the official party. Nevertheless, the local chapter is opposed to the project, and has urged its members to stay in Ojitlan as long as possible.

In Uxpanapa, the local office of the CCI processes individual grievances and puts pressure on the commission to expedite services that have been promised. It plays an important function, but does not represent the interests of its members in opposition to the government. It is inconceivable, for example, that it should ever try to organize a strike to force the commission to abandon the rice program, which has not benefited the overwhelming majority of the ejidatarios. This leaves the people no recourse except passive resistance; they just don't work very hard.

The new towns under construction in Uxpanapa are still fairly rough, but they already have electricity, running water, and other services available in only a few of the ejidos in Ojitlan. Nevertheless, many people have refused to move in until they are entirely completed; partly because they are unhappy about leaving their old homes, partly because their old plots still provide them with a significant portion of their income, and partly as a form of resistance to the program, which has not yet provided them with the benefits they were promised. The lack of an established population greatly complicates both the organization of the labor force in the agricultural program and the organization of the collective ejidos. It is very irritating to the employees of the commission and the bank, who feel that they are working very hard for an ungrateful people who refuse to share the hardships of a newly opened frontier area, and expect to be handed everything on a platter.

Tequios, unpaid cooperative work projects, were organized in Ojitlan to build and maintain municipal facilities such as paths, ejidal buildings, and schools. The custom has not survived the move to Uxpanapa, where the people feel very little obligation to take care of their own needs, in spite of the collective system. They have been very slow to stucco their houses and plant trees and crops in their backyard gardens, and have shown very little enthusiasm for the model garden program of the commission. Young social workers from the cities have been hired to run a conventional home economics extension program with the women. They have quickly become frustrated by the lack of interest and "dirty habits" of the Indians. These are only some of the symptoms of the paternalism pervading the project.

The social structure of the Chinantecs is changing in the new environment. The power of the traditional leaders, caciques, is based on a complex network of personal relationships, both within the community and with the local political elite. They are everybody's compadre, or cogodfather. Many of them make a point of being very Indian; they dress their several women in beautifully woven huipiles, and observe the old customs. Although some of them are almost ostentatiously corrupt, they were able to take care of the needs and problems of their people in the traditional context. Their supremacy is being challenged in Uxpanapa by a very different kind of leader, who is able to bridge the gap between the ejidatarios and the government agencies. These men are literate and understand the complexities of the system being introduced.

One of the greatest frustrations the official staff has faced in organizing the program is that none of them can speak Chinantec. For example, a representative of the bank would go to a meeting and explain some element of the credit system as clearly as possible. The ejidatarios would discuss the proposal at length in their own language, and then vote against it. The bureaucrat would have no way of knowing what it was that they didn't understand or agree with. A literate Chinantec who can translate and facilitate this process can get better and faster service to his people, and is a tremendous help to the overworked staff. He is rewarded by both sides; the ejidatarios vote him into office, and the officials help him to get personal credit for a truck, or look the other way as he sells liquor out of his house in violation of the prohibition regulations in the zone. There are many honest, dedicated, and unselfish leaders among the Chinantecs, but personal influence and wealth are being concentrated by a new elite.

An interesting sociological study could be made of the evolution of this new class structure. It will probably become more pronounced as enterprises such as rubber and sugar cane, which require the division of labor between skilled and unskilled workers, are introduced on a large scale. Figure 10.1 illustrates the stratification of income levels that developed within a few years of the time collective ejidos were set up in the Chontalpa project in Tabasco, which is organized in much the same way as Uxpanapa.

The spontaneous colonists who live in the resettlement zone are in a somewhat different situation, but they have also come into serious conflict with the system being introduced. The commission has cleared land in some of their ejidos by machine, which was represented to them as a free service in return for accepting relocated Chinantec families into the vacancies that were not filled by the original population. In fact, however, there have been strings attached, of which the most important is that the land became part of the official agricultural program. The colonists have had the same problems with the annual crop and credit programs as the Chinantecs. They feel that the income from the timber operations on their land should be paid to them, instead of going into the common fund under the control of the Union of Ejidos, of which they would be minority members if they joined. They may be allowed to set up a separate union of their own.

They must leave the villages which they built for themselves in the wilderness to become eligible for the housing, electricity, and other services they have been offered. There have been a variety of conflicts over boundaries, legal membership lists, and other details. The construction of the road and other infrastructure has opened up many new opportunities to them, but they are impatient with the collective system and the pervasive government control that comes along with it. Most of these ejidatarios would like to increase the area under cultivation and go into livestock production independently of the commission's program.

Many of these problems, which threaten to undermine the Papaloapan Commission's real achievements in Uxpanapa, can be traced back to the fact that the zone as a whole, rather than the individual ejido or the individual family, has been treated as the basic unit of production. The collectives have become identified in the people's minds with a modern agricultural system that has not benefited them.

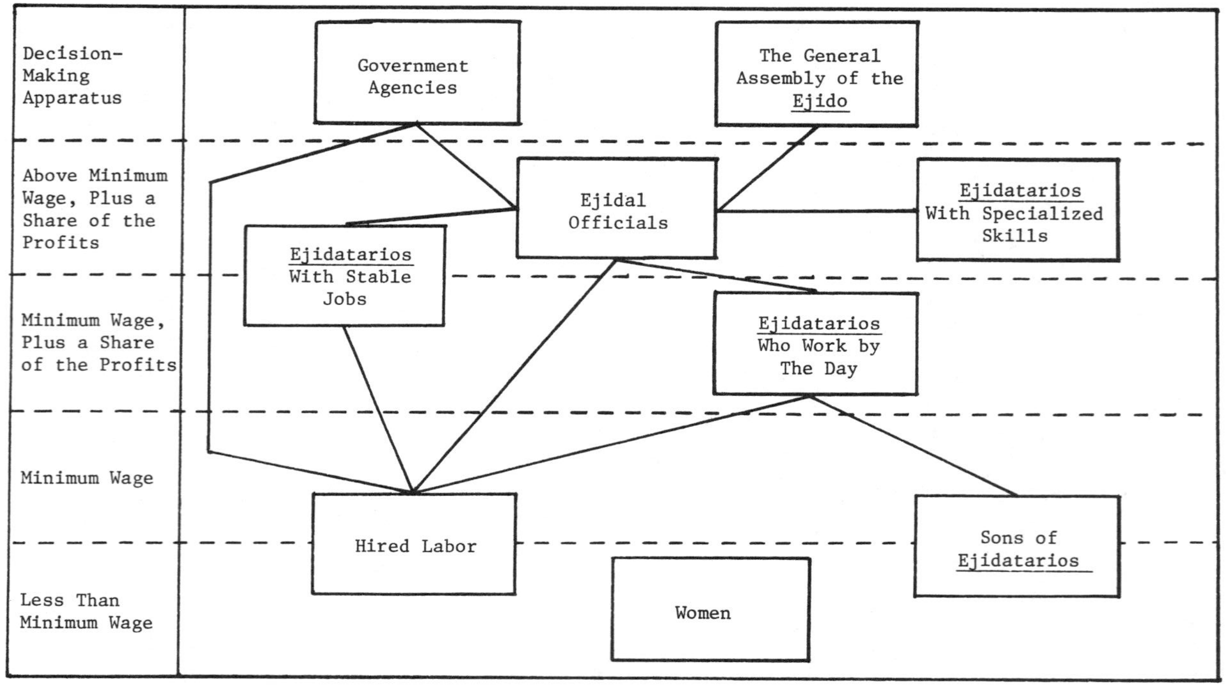

Fig. 10.1. Structure of labor relationships and income levels in a collective ejido.

Source: Adapted from David Baakin, <u>Desarollo Regional y Reorganizacion Campesina</u>, Editorial Neuva Imagen, Mexico, 1978.

They have been numbed by the complexity of the process; the constant meetings and the endless documents they must approve to receive any income at all, but which they do not really understand. They feel like poorly-paid peons on a state farm, and have almost no incentive to work hard. They cannot understand why a person who applies himself should be paid the same as someone who does very little. They do not see how they can build up anything for themselves, and can see a new privileged elite emerging.

It would have been more logical to organize the cooperatives in stages. Teams could have been formed of small groups of people who knew and trusted each other and who had already worked together in the old ejidos. As credit, machinery, and improved technology provided tangible benefits to the members, more and more ejidatarios could have been organized to cultivate larger and larger areas. This was not done, in part because there is a strong prejudice in the country as a whole against the formation of privileged groups.

For years, the official banks established "solidarity groups" and other ad hoc associations within the ejidos to receive credit. This practice, by excluding the majority of the ejidatarios, contributed both to the development of powerful elites and to the fractionalization of the land into small plots (Fernandez y Fernandez, 1965, 1973; Reyes Osorio, 1975). When the relevant laws were revised in the early 1970s to promote modernization in the reformed sector, the ejido itself was made the only legal entity authorized to receive and administer credit.

The laws do permit some flexibility in the formal structure of the ejidos, but the commission and the banking system decided to introduce a uniform, fully collective system into Uxpanapa. There were to be no private assets at all except for the backyard gardens; individuals who already owned cattle would be forced to sell them to the ejidos. This has led to an entirely unnecessary dichotomy in the minds of the ejidatarios between the only two systems they know: the centrally-managed program of the government, and individual plots cultivated using traditional technology. Their experience in the first few years has led them to prefer the latter, and a whole range of intermediate possibilities have been excluded. Collectivization has not been given a political dimension; it has not been presented as a way to combine forces to meet common goals. It is simply a prerequisite for credit. A system that is not productive and does not meet the needs of the people will inevitably evolve into something else; the commission may well lose control over how the ejidos will be organized, precisely because it has tried to exercise too much control.

Uxpanapa is a fertile zone; there is no reason why it should not become productive using some form of modern technology and some kind of cooperative organization. The options available to the Papaloapan Commission are severely circumscribed by a variety of political pressures, by the available technology, and by the legal and institutional structure through which it must implement its policies. The resentment and distrust that have built up in the first years will not be overcome easily. The history of the ejido as a productive form of organization is discouraging, and the record of the Agrarian Reform Ministry in solving these kinds of problems is very poor. Nevertheless, some promising alternatives are being explored.

Functional cooperatives can only be developed if each ejido is organized separately according to the needs and wishes of its members. Although the officials of the commission and the bank have been as flexible as they can, a major reemphasis in the program will be required if the system is to work effectively. This would include the appointment of a well-trained and motivated advisory staff, active training programs, and some kind of moratorium on the unreasonable production targets that continue to be set on an annual basis.

The laws allow each family to cultivate two hectares individually; they should be permitted to do so. Real incentives should be provided to encourage the people to undertake cooperative enterprises on land held in common, but they should not be coerced. It is unfair to tie the income of the ejidatarios to the residual left over after the very high costs of introducing a new and untested technology into the tropics have been deducted. It is counterproductive to pay everybody at the same daily rate, regardless of their skills and productivity.

All of the lines of credit for both annual and perennial crops should be consolidated into a single fund, on the assumption that the ejido will become a profitable entity within a reasonable period, say five years. In the meantime, each family should be paid a regular minimum salary according to the terms of Article 115 of the Rural Credit Law (Mexico, 37), so that they do not have to worry about meeting their basic needs on a week-to-week basis. Separate production teams should be organized for each enterprise, composed of groups of people with appropriate skills and an active willingness to work together. Their incomes should depend on their productivity. Intensive training programs should be established to give the ejidatarios the skills they will need to manage and operate modern agricultural enterprises; literacy classes would be a good place to start. All of the activities in the ejido would be administered by the elected officials under the authority of the General Assembly; they should be paid an adequate salary.

All of these ideas have been expressed by various officials who have worked in Uxpanapa, and all of them are permitted by law. The point is not to replace one rigid system with another; a flexible approach is very necessary. The new cooperatives should start from the immediate interests of the ejidatarios and help them to develop their resources and capabilities in directions consistent with regional and national priorities. There are no economies of scale in an unprofitable enterprise with a poorly motivated and underproductive labor force. Even if the technology is improved, national goals cannot be met through the ejidal structure without the participation of the ejidatarios.

Neither the Chinantecs nor the spontaneous settlers have had any formal way of applying their practical experience in tropical agriculture to the planning, administration, or evaluation of the project. The technical personnel, many of whom have no previous experience in the

Southeast, have made all of the technical decisions. They have made many mistakes of which the ejidatarios were well aware long before the problems became obvious. The people feel so alienated from the experimental programs of the commission that they steal from the test plots, invalidating the data.

In Guatemala a program has been developed that brings the campesinos into close cooperation with both agronomists and social scientists. Traditional varieties and cultural practices are tested side-by-side with various modern packages of improved seeds, fertilization, pest control, and so forth. Careful records are kept of costs, yields, and economic returns. If some change seems promising, the new technology is introduced on a limited commercial scale, and the results are carefully monitored for several years (Hildebrand, 1977). Similar models have been used in other areas (Kassorla, 1977). Some system of this kind in Uxpanapa would not only avoid costly errors, it would involve the ejidatarios in the program so that they would feel a personal stake in the results (Freire, 1975).

The creation of collective ejidos does not guarantee equity. It would be utopian to expect that everybody will be content with the same income, and it will be very difficult to prevent the most profitable enterprises from coming under the control of a small elite. The imposition of a collective structure in other projects in the Southeast, by breaking down the traditional values of community, has accelerated the formation of very unequal patterns of income distribution (Mexico, 3; 37; Valencia de la Rosa). The formal structure of the ejidos is similar superficially to cooperatives in Socialist societies, such as the Chinese Communes, but they cannot function in the same way in a capitalist economy which systematically rewards individual accumulation.

It is difficult to predict what course the Uxpanapa project will take in the future, but the Mexican government will not continue to pour money into a losing operation indefinitely. By 1978 the Papaloapan Commission was already beginning to work outside of the ejidal structure to meet its production targets for the annual crops. In the longer run, it will almost certainly find crops and technological systems that can take advantage of the infrastructure that has been built. Sugar cane is a likely possibility. Some of the more aggressive and able ejidatarios will find opportunities in the new structure and prosper. There is a real danger that the majority will slip back into traditional slash-and-burn agriculture for a few years, supplemented by daily labor in the commercial operations. As the region becomes more fully developed, the range of opportunities open to them will narrow, and they will become permanently marginalized. If, however, the short-term pressures on the project are relaxed, and if a concentrated effort is made, it is still possible that the collective ejidos may become a structure through which the Chinantec people will be able to participate in the modernization of the Southeast.

Appendix — Abridged Translation of the Executive Decree of August 29, 1972, Which Authorized the Construction of the Cerro de Oro Dam

INTRODUCTION

1) The Federal Government, conscious of the need to resolve the social, economic, and technical problems of the watershed basin of the Papaloapan River by taking advantage of the natural resources, not only through the construction of engineering projects, but also by promoting economic development through a variety of integrated programs, created the Papaloapan Commission by executive decree on February 26, 1947. This agency was empowered to plan and to construct all of the works necessary to prevent flooding, develop irrigation systems, build roads, other services, and sanitary systems, and to coordinate industrial, agricultural, and colonization projects.

2) The future socioeconomic development of the nation must be based on a more just and balanced foundation. The distribution of income must be improved in favor of the great majority of the population, particularly in areas which have previously received very few benefits, as is in the case in the upper basin of the Papaloapan. New infrastructure will increase both production and the income of the people, and will permit this region to be integrated into the national economy.

(3-5) (The purpose of the Special Study Commission (1) is explained. The importance of initiating projects in the State of Oaxaca as a tribute to President Benito Juarez is stressed.)

---

(1) The Special Study Commission, composed of Ing. Jorge L. Tamayo, Inc. Victor Bravo Ahuja, and Antonio Jimenex Puyo, was established in May, 1972 at the request of President Echeverria in response to local opposition in the State of Oaxaca to the construction of the Cerro de Oro Dam. Its report was published later in the same year as: Informe de la Junta Especial de Estudios Relativo al Desarrollo Socioeconomico de la Parte Alta de la Cuenca del Rio Papaloapan.

THE DECREE

1) The recommendations and conclusions of the Special Study Commission are hereby approved.

2) The following works will be constructed:

I. A dam to create a storage reservoir with a capacity of 3.5 billion cubic meters will be built at Cerro de Oro on the Santo Domingo River to control flooding and to generate electricity. 70,000 hectares in the State of Oaxaca will be provided with irrigation. A diversion dam will be built on the Valle Nacional River. The capacity of the Temazcal reservoir will be increased to 9.9 billion cubic meters.

II. Complementary works, such as permanent levees and artificial canals, will be built to increase the capacity of the main bed of the Papaloapan River.

III. The soil conservation program in the upper basin will be intensified. One million hectares in the Mixtec zone will receive special attention, with particular emphasis on natural resources such as forests and mineral deposits, roads, and small scale irrigation works.

IV. 363 million pesos are hereby authorized for the construction of roads in that portion of the Papaloapan Basin which lies in the State of Oaxaca.

V. Agricultural experiment stations, animal health centers, and demonstration plots will be established to meet the needs of the region.

(VI-X) (Schools, post and telegraph offices, hospitals, rural health centers, drinking water, sewage, and electrical power will be provided to certain towns, as listed in detail.)

3-4) (The responsibilities of the Papaloapan Commission and other government agencies are specified, and the budget is allocated.)

5) The following compensation will be provided to the population affected by the construction of the Cerro de Oro Dam:

I. <u>Ejidos</u>

a) The ejidos will be relocated by the agrarian authorities, either in the district which will be protected by the dam, or in whatever resettlement zones may be acquired by the Ministry of Hydraulic Resources.

Representatives of each legally constituted ejido or community will select the area into which they wish to be relocated. The Ministry of Hydraulic Resources and the Papaloapan Commission will immediately construct all of the necessary infrastructure. As soon as it is completed, the agrarian authorities will help the people move.

As long as the lakebed is not flooded, the ejidatarios may continue to farm the areas where they currently live, even if they have received land in other areas as compensation.

b) The area which will be granted to each ejidatario or comunero as compensation will be set according to the Agrarian Reform and Water Laws, and will vary depending on whether or not it is provided with irrigation, but in no case will it be smaller than indicated in the appropriate decrees. (2) In addition, each ejido or community will receive at least as much pasture land as it currently uses collectively.

All individuals presently residing in the ejidos and communities, whether or not they currently hold rights, will be included in the census for the purpose of distributing rights in the new areas.

c) The buildings and other assets of the ejidatarios will be compensated in cash on the basis of evaluations carried out by the government. Each ejido will designate representatives to participate in the evaluation process, and their opinions must be taken fully into account.

II. Private Property

a) Private property owners will be compensated in cash according to an evaluation to be carried out by the government. Representatives of the farmers will be selected by the Papaloapan Commission from a list drawn up by the Farmers Association, the Cattle Ranchers Association, and the Association of Private Property Owners of the affected area.

---

(2) It was later decided to grant each ejidatario rights to 20 hectares in Uxpanapa or other rain-fed areas, and ten hectares in irrigated areas. Each new ejido is being organized collectively.

Source: Diario Oficial, Vol. 313, No. 52, August 30, 1972.

b) Each individual may choose between indemnification in cash or in the form of land in other areas.

6) In accordance with the Federal Waters Law, the parcel provided to each private property owner in compensation will not exceed 20 hectares of irrigated land, or 250 hectares of pasture land in areas which are not suitable for crop production.

7) The official credit institutions of the Federal Government will provide all of the campesinos affected by the dam with both long-term and short-term credit to help them to cultivate their lands.

8) The appropriate ministries and decentralized agencies are hereby directed to allocate the financial resources which will be necessary to carry out the provisions of this decree.

[Signed]
Luis Echeverria Alvarez, Constitutional President of the United States of Mexico

References

Aguilar Hernandez, Miguel. 1976. Alcances y Limitaciones de los Nuevos Sujetos de Credito Rural. Tesis Profesional, Facultad de

Aguirre Beltran, Gonzalo. 1957. El Proceso de Aculturacion. Mexico: UNAM.

__________. 1967. Regiones de Refugio. Mexico: Instituto Indigenista Interamericano.

__________. 1976. Ethnocidio en Mexico: Una Denuncia Irresponsable. In B. Gonzalo Aguirre, ed., Obra Polemica. Mexico: SEP-INAH.

Alatorre, Gustavo, Nestor Garay Romo, and Fernando Morales Diaz. 1977. Energia Electrica. In J.L. Tamayo and E. Beltran, eds., Recursos Naturales de la Cuenca del Papaloapan. Mexico: Instituto Mexicano de Recursos Naturales Renovables.

Alcazar, Mario A. 1975. El Plan Chontalpa. In F. Ivan Restrepo, ed., Los Problemas de la Organizacion Campesina. Mexico: Editorial Campesina.

Allen, Elizabeth A. 1978. Infrastructural Investment and New Settlement: The Papaloapan Basin, Mexico. Mimeo, Department of Geography, University of Aberdeen.

Apple, J.L. 1972. Intensified Pest Management Needs of Developing Nations. BioScience, vol. 22, no. 8.

Attolini, Jose. 1949. Economia de la Cuenca del Papaloapan -- Agricultura. Mexico: Instituto de Investigaciones Economicas.

Avila Pacheco, Simon David. 1977. Effectos Sociales de un Proyecto de Desarrollo Regional -- El Caso de la Cuenca del Papaloapan. Tesis Profesional, Facultad de Ciencias Politicas y Sociales, UNAM.

Ballesteros, Juan, et al. 1970. La Colonizacion del Papaloapan. Mexico: Centro de Investigaciones Agrarias.

Banco Nacional de Mexico. 1977. *Review of the Economic Situation of Mexico*, vol. 53, no. 620.

Barabas, Alicia, and M.A. Bartolome. N.d. Informe Antropologico sobre la Poblacion Chinanteca que Sera Afectada por la Presa Cerro de Oro. Mexico: Secretaria de Recursos Hidraulicos, Comision del Paploapan.

__________. 1973. *Hydraulic Development and Ethnocide: The Mazatec and Chinantec People of Oaxaca, Mexico*. Copenhagen: International Work Group for Indigenous Affairs, Document #15.

Barbosa Ramirez, A.R. 1974. *La Ganaderia Privada y Ejidal*. Mexico: Centro de Investigaciones Agrarias.

Barkin, David. 1972. *Los Beneficiarios del Desarrollo Regional*. SepSetentas, 52, Mexico.

__________. 1978. *Desarrollo Regional y Reorganizacion Campesina*. Mexico: Editorial Nueva Imagen.

Barraclough, Solon. 1977. Perspectivas de la Crisis Agricola en America Latina. *Naxi-Nanta, Revista de Economia Campesina*, no. 1 (Mexico).

Barrera, A., A. Gomez-Pompa, and C. Vazques-Yanes. 1977. El Manejo de las Selvas por los Mayas: sus Implicaciones Silvicolas y Agricolas. *Biotica*, vol. 2, no. 2 (Xalapa, Mexico).

Bassols, B.A. 1965. Dos Ejemplos Concretos de Colonizacion del Tropico Mexicano. *El Dia*, May 25 (Mexico).

Benitez, Fernando. 1970. Tierra de Brujos -- El Mundo Echando a Pique. In F. Benitez, *Los Indios de Mexico*. Mexico: Biblioteca Era.

Bevin, Bernard. 1938. *The Chinantec*, Vol. 1. Mexico: Instituto Panamericano de Geografia e Historia.

Blom, P. 1975. Rice Production in the Gulf Coast. Mimeo, Secretaria de Recursos Hidraulicos, Plan Nacional Hidraulico, Mexico.

Bravo Ahuja, Victor, Jorge L. Tamayo, and Antonio Jimenez Puya. 1972. Informe de la Junta Especial de Estudios Relativo al Desarrollo Socio-economico de laParte Alta de la Cuenca del Rio Papaloapan. Mimeo, Mexico.

Brissot, M.A. 1837. *Voyage au Gauzacoalcos, aux Antilles, et aux Etat-Unis*. Paris: Arthus Bertrand.

Caballero Nieto, Javier. 1976. El Costo Ecologico del Uso de la Tierra en un Ejido del Tropico Mexicano. Tesis Profesional, Departamento de Biologia, Facultad de Ciencias, UNAM.

Carrasco Puente, Rafael. 1948. *Bibliografia del Istmo de Tehuantepec* (2301 entries). Mexico: Secretaria de Relaciones Exteriores.

Carrion Spindola, C.M., J.L. Martinez Cruz and A. Ortiz Cruz. 1977. Plan Uxpanapa. Tesis Profesional, Facultad de Arquitectura, Universidad Veracruzana.

Casteneda Guzman, Luis. 1958. Cabeza de Puente Yanqui en Tehuantepec. Mexico: Editorial Jus.

Centro Internacional de Agricultura Tropical (CIAT). 1975. Horizontal Resistance to the Blast Disease in Rice. Cali, Colombia.

__________. 1977a. Annual Report, 1976. Cali, Colombia.

__________. 1977b. Workshop on Strategies to Improve Rice Production in Latin America. Cali, Colombia.

Centro Internacional de Mejoramiento de Maiz y Trigo (CIMMYT). 1974. El Plan Puebla -- Siete Anos de Experiencia. Mexico.

Chao Ebergenyi, Guillermo. N.d. El Reto de la Chontalpa. Mexico: Villahermosa.

Chayanov, A.V. 1966. The Theory of Peasant Economy. Homewood, Ill.: The American Economic Association.

Cheaney, Robert L., and Peter R. Jennings. 1975. Field Problems of Rice in Latin America. Cali, Colombia: Centro Internacional de Agricultura Tropical.

Cisneros Vasquez, Rafael, and Xavier Gamboa Villafranca. 1973. Informe Final de la Investigacion de Antecedentes del Reacomodo Para Implementar Futuras Acciones -- El Caso de la Presa Presidente Aleman. Mimeo, Secretaria de Recursos Hidraulicos, Comision de Relaciones Humanas, Mexico.

Clark, Colin, and Margaret Haswell. 1967. The Economics of Subsistence Agriculture. New York: St. Martin's Press.

Cline, Howard F. 1957. Problems of Mexican Ethno-History: The Ancient Chinantla. The Hispanic American Historical Review, vol. 37, no. 3.

Coe, Michael D. 1962. Mexico. New York: Praeger.

__________. 1966. The Maya. New York: Praeger.

__________, and Kent V. Flannery. 1967. Early Cultures and Human Ecology in South Coastal Guatemala. Washington, D.C.: Smithsonian Press.

Conklin, Harold C. 1963. The Study of Shifting Cultivation. Washington, D.C.: The Panamerican Union.

Cook, Sherburne, and Lesley Bird Simpson. 1948. The Population of Central Mexico in the Sixteenth Century. In Ibero-Americano. Vol. 31. Berkeley: University of California Press.

Covarrubias, Miguel. 1948. Mexico South -- The Isthmus of Tehuantepec. New York: Knopf.

Dasman, Raymond F., John P. Milton, and Peter H. Freeman. 1973. Ecological Principles for Economic Development. London: Wiley.

DelCorral, Miguel. 1777. La Costa de Sotovanto. Mexico: Editorial Citaltepetl, reprinted 1964.

Delgado, Augustin. 1966. Arqueologia de la Chinantla, Noreste de Oaxaca: su Secuencia Actual. In Summa Antropoligica en Homenaje a Roberto J. Weitlaner. Mexico: INAH.

Diaz del Castillo, Bernal. 1939. Historia Verdadera de la Conquista de la Nueva Espana. Mexico: Editorial Pedro Robredo.

Dorozynski, Alexander. 1978. Wanted: A Sustainable System for the Tropics. Ceres, vol. 11, no. 4.

Drumond, D.E. 1968. Swidden Agriculture and the Rise of Maya Civilization. Southwestern Journal of Anthropology, vol. 17.

Duran, Marco Antonio. 1975. El Agrarismo Mexicano. 3rd ed. Mexico: Siglo XXI.

Echenique-Manrique, Ramon. 1975. Estudio Botanico y Ecologico de la Region del Rio Uxpanapa, Veracruz: Caracteristicas Tecnologicas de la Madera de Diez Especies. Mexico: CONACYT, Programa Nacional Indicativo de Ecologia Tropical, INIREB.

__________. 1976. Durmientes del Metro. Mexico: INIREB, Comunicado no. 4.

Eckstein, Salomon. 1966. El Ejido Colectivo en Mexico. Mexico: Fondo de Cultura Economica.

Escanero Gonzalez, Francisco. 1977. Reacomodo de Campesinos Afectados por la Construccion de la Presa Cerro de Oro en Uxpanapa, Veracruz. Tesis Profesional, Facultad de Ingenieria, UNAM.

Espinosa, Mariano. 1910. Apuntes Historicos de las Tribus Chinantecas, Mazatecas, y Populucas. Reprinted 1961 with notes by H.F. Cline. Papeles de la Chinantla no. 7. Mexico: INAH.

Far Eastern Economic Review. 1977. Forests, Save or Squander? Vol. 98, no. 48.

Fernandez y Fernandez, Ramon. 1965. Economia Agricola y Reforma Agraria. Mexico: Centro de Estudios Monetarios Latinoamericanos.

__________. 1973. Cooperacion Agricola y Organizacion Economica del Ejido. SepSetentas, Mexico.

__________. 1975. Perspectivas del Ejido. Chapingo, Mexico: Colegio de Postgraduados, Escuela Nacional de Agricultura.

__________. 1978. El Problema de los Alimentos y la Tenancia de la Tierra. Paper presented at Symposium on "Food, a Challenge to Mexico," Jurica, Mexico.

Freire, Paulo. 1975. Extension o Comunicacion? La Concientizacion en el Medio Rural. Mexico: Siglo XXI.

Gerhard, Peter. 1972. A Guide to the Historical Geography of New Spain. Cambridge, Mass.: Cambridge University Press.

Glanz,Susana. 1974. El Ejido Colectivo de Nueva Italia. Mexico: INAH.

Glick, Edward B. 1953. The Tehuantepec Railroad: Mexico's White Elephant. Pacific Historical Review, vol. 22, no. 4.

Gomez Jara, Francisco. 1970. El Movimiento Campesino en Mexico. Mexico: Editorial Campesina.

Gomez-Pompa, Arturo. 1971. Las Regiones Tropicales de Mexico y el Aprovechamiento de sus Recursos. Boletin de Divulgacion, no. 6, Sociedad Mexicana de Historia Natural, Mexico.

__________, C. Vasquez-Yanes, and S. Guevara. 1972. The Tropical Rain Forest -- A Non-renewable Resource. Science, vol. 177, no. 4051.

__________, and Antonio Lot, eds. 1974. Segundo Informe del Estudio Botanico y Ecologico de la Region del Rio Uxpanapa, Veracruz. Mimeo, Instituto de Biologia, UNAM.

__________. 1975. Prologue to Estudio Botanico y Ecologico de la Region del Rio Uxpanapa, Veracruz: Caracteristicas Tecnologicas de la Madera de Diez Especies, by Ramon Echenique-Manrique. Mexico: CONACYT, Programa Nacional Indicativo de Ecologia Tropical, INIREB.

Gomez Tagle, Silvia. 1974. Organizacion de las Sociedades de Credito Ejidal de la Laguna. Mexico: Centro de Estudios Sociologicos, El Colegio de Mexico.

Gordon, David. 1978. Country Survey #3 -- Mexico. The Economist, vol. 267, no. 7025 (London).

Gourou, Pierre. 1966. The Tropical World. 4th ed. New York: Wiley.

Greenland, D.J. 1975. Bringing the Green Revolution to the Shifting Cultivator. Science, vol. 190, no. 4217.

Grist, D.H. 1975. Rice. 5th ed. London: Longmans.

Guerere, Alberto, and Efrain Hernandez X. 1970. Uso de la Tierra en la Region de Tuxtepec, Oaxaca. Publicacion Especial no. 6, Instituto Nacional de Investigaciones Forestales, Mexico.

Hansen, Roger D. 1971. The Politics of Mexican Development. Baltimore: Johns Hopkins Press.

Hernandez Aragon, Leonardo. 1976. Arrox. Mimeo, Secretaria de Agricultura y Ganaderia, Instituto Nacional de Investigaciones Agricolas, Programs de Mejoramiento de Arroz, Mexico.

Hernandez X., Efrain. 1959. La Agricultura. In J.L. Tamayo and E. Beltran, eds., Los Recursos Naturales del Sureste y su Aprovechamiento. Mexico: Instituto Mexicano de Recursos Naturales Renovables.

Hewitt de Alcantara, Cynthia. 1976. Modernizing Mexican Agriculture. Geneva: United Nations Research Institute for Social Development.

_________. 1977. Ensayo Sobre la Satisfaccion de Necesidades Basicas del Pueblo Mexicano entre 1940 y 1970. Mexico: Centro de Estudios Sociologicos, El Colegio de Mexico.

Higuera Ciapara, Inocencio. 1977. Alimentos, Poblacion, y Empleo. Ithaca, N.Y.: Cornell Agricultural Economics Staff Paper 77-1.

Hildebrand, Peter E. 1977. Generando Technologia para Agricultores Tradicionales: Una Metodologia Multidisciplinaria. Guatemala, Sector Publico Agricola, Instituto de Ciencia y Tecnologia Agricolas (ICTA).

Huey, John. 1978. No Fiesta Ahead. The Wall Street Journal, August 30.

vonHumboldt, Alexander. 1811. Political Essay on the Kingdom of New Spain. London: Longman, Hurst, Rees, Orme, and Brown.

Iglesias, Jose Maria. 1831. Acayucan. Mexico: Editorial Citaltepetl, reprinted 1966.

International Rice Research Institute (IRRI). 1965. The Rice Blast Disease. Baltimore: Johns Hopkins University Press.

_________. 1975. Major Research in Upland Rice. Los Banos, The Philippines.

International Union for the Conservation of Nature and Natural Resources. 1975. The Use of Ecological Guidelines for Development in the American Humid Tropics. Morges, Switzerland.

Janzen, Daniel H. 1973. Tropical Agroecosystems. Science, vol. 182, no. 4118.

Kassorla, Jackie. 1977. Beginning Agricultural Research with the Farmer. M.P.S. Thesis, Cornell University, Ithaca, N.Y.

Lewis, W. Arthur. 1964. Thoughts on Land Settlement. In Carol Eicher and Lawrence Witt, eds., Agriculture in Economic Development. New York: McGraw-Hill.

Lopez Arredondo, Gonzalo. 1976. Enfermedades que Atacan al Arroz en las Zonas Temporaleras. Mimeo, Secretaria de Agricultura y Ganaderia, Instituto Nacional de Investigaciones Agricolas, Campo Experimental del Istmo de Tehuantepec, Mexico.

Lopez Portillo, Jose, President of Mexico. 1978. State of the Union Message, September.

Lucero, Alvaro F., and Simon David Avila. 1974. Proyecto Cerro de Oro. Manuscript, Mexico.

_________. 1977. Alternativas para el Reacomodo de la Poblacion Afectada por la Construccion de la Presa Cerro de Oro. Mimeo, Mexico.

_________. 1978. Efectos Sociales y Ecologicos de los Programas de Desarrollo en la Cuenca del Rio Papaloapan. Paper presented at the annual meeting of the Society for Applied Anthropology, Merida, Mexico.

McBride, George M. 1923. <u>The Land Systems of Mexico</u>. New York: American Geographical Society.

McDowell, R.E. 1972. <u>Improvement of Livestock Production in Warm Climates</u>. San Francisco: W.H. Freeman.

McMahon, David F. 1973. <u>Antropologia de Una Presa: Los Mazatecos y el Proyecto del Papaloapan</u>. Mexico: Instituto Nacional Indigenista.

Marquez Cabrera, Fernando, and Manuel Valenzuela C. 1976. <u>Macuspana A-75</u>; Nueva Variedad de Arroz para las Zonas Temporaleras de Tabasco y Regiones Similares del Sureste. Cardenas, Tabasco, Mexico: Colegio Superior de Agricultura Tropical.

Martinez Alfaro, Miguel. 1970. Ecologia Humana del Ejido Benito Juarez o Sebastopol, Tuxtepec, Oaxaca. Boletin Especial del Instituto Nacional de Investigacion Forestal, no. 7, Mexico.

Matheny, Ray T. 1977. Maya Lowland Hydraulic Systems. <u>Science</u>, vol. 193, no. 4254.

Melgarejo Vivanco, Jose Luis. 1960. <u>Breve Historia de Veracruz</u>. Xalapa, Mexico: Universidad Veracruzana.

<u>Mexico</u> -- documents and publications of agencies of the Mexican Government without author citations are listed in the following section.

1. Banco de Credito Rural del Golfo, documents on file in the regional office in Acayucan, Veracruz, and in the local branch in La Laguna, Uxpanapa (Veracruz).
2. Comision Coordinadora para el Desarrollo Integral del Istmo of Tehuantepec. N.d. <u>Invest in the Isthmus of Tehuantepec</u>.
3. CONACYT, Centro de Ecodesarrollo. 1976. Evaluacion del Plan Chontalpa.
4. Departamento de Asuntos Agrarios y Colonizacion (DAAC). 1971. <u>Informe de Labores</u>.
5. Direccion General de Estadistica. 1973. <u>Noveno Censo General de Poblacion -- 1970</u>.

6. Fideicomiso de Hule, Programa Nacional de Hule. Data on file in office in La Laguna, Uxpanapa (Veracruz).

7. Fideicomiso Para el Aprovechamiento de la Madera Utilizable en Uxpanapa. Data on file in office in Matias Romero, Oaxaca.

8. Fondo Nacional de Fomento Ejidal, Gerencia de Proyectos. 1975. Desarrollo Regional -- Uxpanapa, Veracruz. Mimeo.

9. Secretaria de Agricultura y Recursos Hidraulicos, Centro de Investigaciones Agricolas de Sinaloa. 1977. Cultivo de Arroz en el Valle de Culiacan. Circular CIAS no. 8, 6a Edicion.

10. __________, Campo Agricola Experimental del Istmo de Tehuantepec. 1977. Informe Annual -- 1976. Mimeo.

11. __________, Comision del Papaloapan, Agrologia Desarrollo, S.C. 1977. Estudio Semi-detallado -- Distrito de Drenaje de Uxpanapa. Mimeo.

12. __________, Comision del Papaloapan. 1977. Breve Resena de la Obra de Uxpanapa. Mimeo.

13. __________, Comision del Papaloapan. 1978. Recomendaciones Tecnicas para las Actividades Agropecuarias. Mimeo.

14. Secretaria de Recursos Hidraulicos, Comision del Papaloapan, Comite de Reacomodo, Presa Cerro de Oro. Records and documents on file in Ojitlan, Oaxaca, and in La Raya, Uxpanapa (Veracruz).

15. __________. Boletin Informativo, various issues.

16. __________. 1962. Urgencia de Iniciar la Construccion de la Presa Cerro de Oro. Mimeo.

17. __________. 1973. Boletin Hidrometrico, no. 20. Cd. Aleman, Veracruz.

18. __________. CIEPS S.C. 1973. Rio Santo Domingo; Estudio de Factabilidad Tecnica, Economica, Social, y Financiera. Mimeo.

19. __________. COPLASA. 1973. Investigacion Socio-economica de la Vivienda en el Area de la Presa Cerro de Oro. Mimeo.

20. __________. 1974. El Ganado Bovino en la Cuenca Baja del Rio Papaloapan y Las Caracteristicas de Explotacion y Manejo. Mimeo.

21. __________, Estudios y Proyectos, S.A. 1975. Reconocimiento Agrologico Para el Desarrollo Agropecuario de Uxpanapa. Mimeo.

22. __________. 1974. Atlas Climatologico e Hidrologico de la Cuenca del Papaloapan.

23. __________. 1976. Memoria de la Construccion de la Obra en Uxpanapa. Mimeo.

24. __________, Direccion del Manejo de Cuencas. 1976. Mesa Redonda sobre Manejo de Cuencas. Mimeo.

25. _________, Comite de Reacomodo, Presa Cerro de Oro. 1976. Tercer Informe. Mimeo.

26. Secretaria de Recursos Hidraulicos, Residencia de Estudios Sociales, Comision de Relaciones Humanas, archives.

27. _________, Subsecretaria de Planeacion, Estudios y Proyectos, S.A. 1972. Estudio Agrologico Especial del Vaso de Cerro de Oro y las Areas de Ojitlan y San Felipe Usila, Oaxaca. Mimeo.

28. _________, Subsecretaria de Planeacion. 1976. Estudio Sobre el Proyecto para el Control del Rio Papaloapan. Mimeo.

29. _________, Plan Nacional Hidraulico. 1975. Plan Nacional Hidraulico.

30. _________, Plan Nacional Hidraulico. 1975. Diagnostico de la Migracion en Mexico. Mimeo.

31. _________, Plan Nacional Hidraulico. 1976. Proyecto Piloto para el Desarrollo de la Planicie Costera del Golfo de Mexico -- Estudio Socioeconomico, Tacotalpa. Mimeo.

32. _________, Plan Nacional Hidraulico. 1977. Plan Piloto Tesechoacan -- Diagnostico Socioeconomico. Mimeo.

33. Secretaria de la Reforma Agraria. Records and documents on file in La Laguna, Uxpanapa (Veracruz).

34. Law. 1961. Ley y Reglamento del Seguro Agricola, Integral y Ganadero. Diario Oficial. December 30.

35. Law. 1971. Ley Federal de Reforma Agraria. Diario Oficial. April 16.

36. Law. 1972. Ley Federal de Aguas. Diario Oficial. January 11.

37. Law. 1976. Ley General de Credito Rural. Diario Oficial. May 26.

38. Executive Decree. 1972. Decreto por el que se aprueba el dictamen y las conclusiones de la Junta Especial de Estudios, relativas al desarrollo socioeconomico de la parte alta de la cuenca del Rio Papaloapan. Diario Oficial. August 30.

39. Executive Decree. 1974. Decreto que declara de utilidad publica el establecimiento del Distrito de Drenaje de Uxpanapa. Diario Oficial. January 14.

Moraes, Vicente. 1977. Rubber. In Paulo de T. Alvim and T.T. Kozlowski, eds., Ecophysiology of Tropical Crops. New York: Academic Press.

National Academy of Sciences. 1975. Underexploited Tropical Plants with Promising Economic Value. Mimeo.

Nelson, Michael. 1973. The Development of Tropical Lands: Policy Issues in Latin America. Resources for the Future. Baltimore: Johns Hopkins University Press.

New York Times. 1978. November 29.

Noriega R., Jose. 1946. Control del Rio Papaloapan: Preparacion del Plan de Estudios Definitivos y Programa de Construccion de las Obras. Reprinted in 1973 by la Comision del Papaloapan, Comision Nacional de Irrigacion, Mexico.

Ortiz de Ayala, Tadeo. 1831. Istmo de Tehuantepec. Mexico: Editorial Citlatepetl, reprinted 1966.

Ortiz Wadgymar, Arturo. 1971. Aspectos de la Economia del Istmo de Tehuantepec. Mexico: Instituto de Investigaciones Economicas, UNAM.

Osorio A., Mario. N.d. La Ganaderia Bovina en el Estado de Tabasco. Cardenas, Tabasco: Colegio Superior de Agricultura Tropical. Mimeo.

Ou, S.H. 1972. Rice Diseases. Kew, England: Commonwealth Mycological Institute.

Palma Gomez, Adalberto. 1977. Substitucion de Importaciones de Productos Agropecuarios Alimenticios en Mexico; Analisis de Posibilidades; el Caso de Grasas y Aceites. Ithaca, N.Y.: Cornell Agricultural Economics Staff Paper no. 77-7.

Pearson, Charles, and Anthony Pryor. 1978. Environment: North and South. New York: Wiley Interscience.

Pepper, Special Agent. 1906. The Isthmus of Tehuantepec and Its Inter-ocean Railway. Scientific American Supplement, vol. 61, no. 1579.

Plancarte Mendez, Ignacio, and Mario A. Alcaraz. 1977. El Programa de Hule en Uxpanapa. Mimeo, Banco de Mexico, FIRA.

Poleman, Thomas T. 1964. The Papaloapan Project. Palo Alto, Calif.: Stanford University Press.

__________. 1978. Food, Population, and Employment: Some Implications for Mexico's Development. Ithaca, N.Y.: Cornell Agricultural Economics Staff Paper no. 78-19.

Polhamus, Loren G. 1962. Rubber. New York: Wiley Interscience.

Purseglove, J.W. 1975a. Tropical Crops -- Monocotyledons. New York: Wiley.

__________. 1975b. Tropical Crops -- Dicotyledons. London: Longman Group.

Rankel, Arthur J. 1955. Ritual Relationships in Ojitlan, Mexico. American Anthropologist, vol. 57, no. 5.

Rappaport, Roy A. 1971. The Flow of Energy in an Agricultural Society. Scientific American, vol. 224, no. 3.

Rendon, Silvia P. 1973. Informe Sobre la Investigacion "Estructura de Poder" en la Zona de Afectacion de la Presa Cerro de Oro. Secretaria de Recursos Hidraulicos, Comision de Relaciones Humanas. Mimeo.

Restrepo F., Ivan, ed. 1975. Los Problemas de la Organizacion Campesina. Mexico: Editorial Campesina.

__________, and Salomon Eckstein. 1975. La Agricultura Colectiva en Mexico -- La Experiencia de la Laguna. Mexico: Siglo XXI.

Revel-Mouroz, Jean. 1972. Amenagement et Colonisation du Tropique Humide Mexicain; le Versant du Golfe et des Caraibes. Paris: Instutut des Hautes Etudes de L'Amerique Latine. Mimeo.

Reyes Osorio, Sergio, Rodolfo Stavenhagen, Salomon Eckstein, Juan Ballesteros, and collaborators. 1974. Estructura Agraria y Desarrollo Agricola en Mexico. Mexico: Centro de Investigaciones Agrarias, Fondo de Cultura Economica.

Reyes Osorio, Sergio. 1975. Hacia una Politica de Organizacion Economica en el Sector Rural. In Ivan Restrepo F., ed., Los Problemas de la Organizacion Campesina. Mexico: Editorial Campesina.

Reynolds, Clark W. 1970. The Mexican Economy. New Haven, Conn.: Yale University Press.

Richards, Paul W. 1973. The Tropical Rain Forest. Scientific American, vol. 229, no. 6.

Rivera, Miguel Angel. 1977. Uxpanapa: Ethnocidio y Ecocidio. Proceso, Ano 1, no. 30 (Mexico City).

Rodriguez Cisneros, Manuel, et al. 1970. Caracteristicas de la Agricultura Mexicana. Mexico.

Rodriguez Vallejo, Jose, and Hector Duarte Penalosa. 1975. El Hule Hevea en la Cuenca del Rio Papaloapan: Antecedentes y Bases para un Programa de Hule en Los Valles del Uxpanapa y Bajo Papaloapan. Secretaria de Recursos Hidraulicos, Comision del Papaloapan, Mexico.

Rodriguez Vallejo, Jose. 1976. Ixcatl, El Algodon Mexicano. Mexico: Fondo de Cultura Economica.

__________. 1977a. La Agricultura en la Cuenca del Papaloapan. In Jorge L. Tamayo and Enrique Beltran, eds., Recursos Naturales de la Cuenca del Papaloapan. Mexico: Instituto Mexicano de Recursos Naturales Renovables.

__________. 1977b. Unpublished study on the Papaloapan District.

Romanini, Claudio. 1976. Ecotenicas para el Tropico Humedo. Mexico: CONACYT, Centro de Ecodesarrollo, Programa de Naciones Unidas para el Medio Ambiente.

Romero, Matias. 1898. Coffee and India-Rubber Culture in Mexico. New York: G.P. Putnam.

Rosales Marrufo, Rafael F. 1977. Hidrologia. In Jorge L. Tamayo and Enrique Beltran, eds., Recursos Naturales de la Cuenca del Papaloapan. Mexico: Instituto Mexicano de Recursos Naturales Renovables.

Ruthenberg, Hans. 1976. Farming Systems in the Tropics. Oxford: Clarendon Press.

Sanchez, Pedro, and S.W. Buol. 1975. Soils of the Tropics and the World Food Crisis. Science, vol. 188, no. 4188.

Sanchez, Pedro. 1976. Properties and Management of the Soils in the Tropics. New York: Wiley.

Scherr, Sara J. 1978. Development, Employment and Equity in Mexico: The Case of the Papaloapan Basin. M.S. thesis, Cornell University, Ithaca, N.Y.

Scobie, Grant M., and Rafael Posada T. 1977. The Impact of High-yielding Rice Varieties in Latin America with Special Emphasis on Colombia. Cali, Colombia: Centro Internacional de Agricultura Tropical (CIAT).

Shepherd, R. 1975. Rubber Planting in the Gulf Coast Region of Mexico. Secretaria de Recursos Hidraulicos, Plan Nacional Hidraulico, Mexico.

Siemens, Alfred Henry. 1964. The Character and Recent Development of Agricultural Settlement in Southern Veracruz, Mexico. Ph.D. diss., University of Wisconsin, Madison.

Silos, Jose S., et al. 1975. El Sector Agricola: Comportamiento y Estrategia de Desarrollo. Comision Nacional Coordinadora del Sector Agropecuario, Nota Technica no. 6, Mexico.

Silva Herzog, Jesus. 1959. El Agrarismo Mexicano y la Reforma Agraria. Mexico: Fondo de Cultra Economica.

Simpson, Eyler N. 1937. The Ejido: Mexico's Way Out. Chapel Hill: University of North Carolina Press.

Stavenhagen, Rodolfo. 1969. A Land Reform Should Answer the Questions it Raises. Ceres, vol. 2, no. 6.

__________. 1970. Social Aspects of Agrarian Reform in Mexico. In Rodolfo Stavenhagen, ed., Agrarian Problems and Peasant Movements in Latin America. New York: Anchor Books.

__________. 1976. Capitalismo y Campesinado en Mexico. Mexico: INAH.

Staver, Charles, 1977. A Generalized Framework for the On-Farm Evaluation of Pasture Productivity and Utilization. M.P.S. Project, Cornell University, Ithaca, N.Y.

Tamayo, Jorge L. 1964. El Problema Fundamental de la Agricultura Mexicana. Mexico: Instituto Mexicano de Investigaciones Economicas.

_________. 1976a. Una Experiencia Forestal Industrial. Mexico Agrario, Ano 4, no. 2.

_________. 1976b. Speech delivered at the Universidad Veracruzana, Xalapa, Mexico, November 29.

_________, and Leandro Rovirosa Wade. 1974. Instrucciones Sobre el Reacomodo en la Presa Cerro de Oro. El Nacional, October 11 (Mexico City).

_________, and Enrique Beltran. 1977. Recursos Naturales de la Cuenca del Papaloapan. Mexico: Instituto Mexicano de Recursos Naturales Renovables.

Thompson, Eric S. 1954. The Rise and Fall of Maya Civilization. Norman: University of Oklahoma Press.

Thurston, H.D. 1975. Tropical Plant Pathology. Ithaca, N.Y.: Department of Plant Pathology, Cornell University.

Toledo, Victor Manuel, Arturo Arguenta, Patricia Rojas, Cristina Mapes, and Javier Caballero. N.d. Uso Multiple del Ecosistema, Conocimiento Tradicional, y Estrategias de Ecodesarrollo. Mexico: Instituto de Biologia, UNAM.

Turner, John Kenneth. 1969. Barbarous Mexico. Austin: University of Texas Press.

Unasylva (FAO Forestry Journal). 1976. Management and Utilization of the Tropical Moist Forest, vol. 28, nos. 112-113.

Universidad Veracruzana, Instituto de Antropoligia. 1973. Indicatores Socioeconomicas de los Municipios Veracruzanos. Xalapa, Mexico.

Valencia de la Rosa, Ruth. N.d. Surgimiento de Diferencias de Clase en los Ejidos de la Chontalpa. Colegio de Postgraduados, Escuela Nacional de Agricultura, Chapingo, Mexico.

Vicente-Chandler, Jose. 1974. Intensive Grassland Management in the Humid Tropics of Puerto Rico. Rio Piedras: University of Puerto Rico Experiment Station Bulletin no. 233.

Villa Rojas, Alfonso. 1955. Los Mazatecos y el Problema Indigena de la Cuenca del Papaloapan. Mexico: Instituto Nacional Indigenista.

Warman, Arturo. 1972. Los Campesinos, Hijos Predilectos del Regimen. Mexico: Editorial Nuestro Tiempo.

_________. 1976. ...Y Venimos a Contradecir. Mexico: INAH.

Watters, R.F. 1968. La Agricultura Migratoria en Mexico. Merida, Venezuela: Instituto Forestal Interamericano de Investigacion y Capacitacion.

Weitlaner, Irmegard. 1936. A Chinantec Calendar. American Anthropologist, vol. 38, no. 2.

Weitlaner, Roberto J. 1951. Notes on the Social Organization of Ojitlan, Oaxaca. Homenaje a Don Alfonso Caso. Mexico: Nuevo Mundo.

_________, and Carlo Antonio Castro G. 1954. Mayultianguis y Tlacoatzintepec. Mexico: Papeles de la Chinantla I, INAH.

_________. 1973. Usila (Morada de Colibries). Mexico: Papeles de la Chinantla VII, INAH.

Weitlaner, Roberto J., and Howard P. Cline. 1969. The Chinantec. In Robert Wauchope, ed., Handbook of Middle American Indians. Vol. 7. Austin: University of Texas Press.

Weitlaner, Roberto J., and Irmegard Weitlaner. 1946. The Mazatec Calendar. American Antiquity, vol. 11, no. 3.

Wellhausen, Edwin J. 1976. The Agriculture of Mexico. Scientific American, vol. 235, no. 3.

Whetten, Nathan L. 1948. Rural Mexico. Chicago: University of Chicago Press.

Wilkie, James W. 1967. The Mexican Revolution, Federal Expenditure and Social Change Since 1910. Berkeley: University of California Press.

Williamson, C., and W.J.A. Payne. 1965. An Introduction to Animal Husbandry in the Tropics. London: Longmans.

Womack, John. 1968. Zapata and the Mexican Revolution. New York: Knopf.

Index

About the Authors

PETER T. EWELL was born in New Haven, Connecticut, and studied at Yale and Cornell Universities. Having worked in Chile and traveled extensively, he has been interested in agricultural development in Latin America for the past ten years. He currently lives in Ithaca, New York.

THOMAS T. POLEMAN is professor of international food economics at Cornell University. He received his B.S. in agriculture from the University of Missouri and did graduate work in agricultural economics at the same university; his doctorate in international agricultural economics is from Stanford University.

Since 1946 Dr. Poleman has completed many research studies and consulting assignments in Mexico, the Caribbean, West Africa, East Africa, Mauritius, and South and Southeast Asia. He has served as a consultant to several organizations, including the FAO, the Rockefeller Foundation, and the U.S. Department of Agriculture.

His research projects at Cornell University cover employment and welfare in developing countries, the rise and stabilization of human populations, and the economics of tropical agricultural products.

Dr. Poleman's publications include three books on the economics of food and agriculture in the tropics, and a number of monographs, articles, and chapters in contributed works in this field.